Lecture Notes in Artificial Intelligence   9046

Subseries of Lecture Notes in Computer Science

LNAI Series Editors

Randy Goebel
 University of Alberta, Edmonton, Canada
Yuzuru Tanaka
 Hokkaido University, Sapporo, Japan
Wolfgang Wahlster
 DFKI and Saarland University, Saarbrücken, Germany

LNAI Founding Series Editor

Joerg Siekmann
 DFKI and Saarland University, Saarbrücken, Germany

More information about this series at http://www.springer.com/series/1244

Jesse Davis · Jan Ramon (Eds.)

Inductive Logic Programming

24th International Conference, ILP 2014
Nancy, France, September 14–16, 2014
Revised Selected Papers

Editors
Jesse Davis
Department of Computer Science
KU Leuven
Leuven
Belgium

Jan Ramon
Department of Computer Science
KU Leuven
Leuven
Belgium

ISSN 0302-9743 ISSN 1611-3349 (electronic)
Lecture Notes in Artificial Intelligence
ISBN 978-3-319-23707-7 ISBN 978-3-319-23708-4 (eBook)
DOI 10.1007/978-3-319-23708-4

Library of Congress Control Number: 2015950031

LNCS Sublibrary: SL7 – Artificial Intelligence

Springer Cham Heidelberg New York Dordrecht London

Printed on acid-free paper

Springer International Publishing AG Switzerland is part of Springer Science+Business Media
(www.springer.com)

Preface

This volume contains revised versions of papers from the 24th International Conference on Inductive Logic Programming. All papers that appear in these proceedings were reviewed by at least three members of the ILP 2014 Program Committee. The ILP conference was held from September 14th to 16th, 2014, in Nancy, France and was co-located with ECML/PKDD 2014. Since its inception in 1991, the annual ILP conference has served as the premier international forum on learning from structured relational data. The conference covers topics such as the induction of logic programs, learning from data represented with logic, multi-relational machine learning, learning from graphs, and applications of these techniques to important problems in fields like bioinformatics, medicine, and text mining.

Prior to the conference, ILP 2014 solicited three types of paper submissions: long papers, short work-in-progress papers, and short papers detailing previously published research. The conference received 17 long-paper submissions of which nine were accepted to appear in these proceedings. The authors of one of these papers elected to submit their work to the Machine Learning Journal special issue on Inductive Logic Programming. Hence, this paper is not included in these proceedings. Five of the remaining long-paper submissions were presented as work-in-progress papers at the conference. We received 24 short paper submissions, which were initially selected by the Program Committee chairs based on relevance. Twenty-one of these papers were presented at the conference. Of these, five were papers that highlighted work relevant to ILP that were recently published in another conference proceedings volume or journal.

Following the conference, 13 of the work-in-progress papers were invited to submit an extended version of their paper submission, which would be reviewed by the ILP Program Committee with accepted papers appearing in this LNCS volume. Eight authors submitted extended papers of which six were accepted.

The conference also featured three invited talks. Helmut Simonis from the University of Cork gave a talk entitled "Learning Constraint Models from Example Solutions." His talk presented ModelSeeker, which is a tool that is able to generate constraint programs based on only observing a small number of model solutions. Sumit Gulwani from Microsoft Research gave a talk entitled "Applications of Program Synthesis to End-User Programming and Intelligent Tutoring Systems." His talk discussed his ongoing work on program synthesis, and highlighted the ability to induce programs from very few training examples. Finally, Lise Getoor from the University of California - Santa Cruz gave a talk entitled "Scalable Collective Reasoning Using Probabilistic Soft Logic." She talked about the theory underlying probabilistic soft logic and highlighted how this framework could very quickly solve large collective reasoning problems. Lise's talk was a joint invited talk with ECML/PKDD, which was generously sponsored by the Artificial Intelligence Journal.

At ILP 2014, the Machine Learning Journal generously continued its sponsorship of the ILP best student paper award. This year, the paper by Andrew Cropper and

Stephen Muggleton titled "Logical Minimisation of Metarules in Meta-Interpretive Learning" received the best paper award and the work by Ryosuke Kojima and Taisuke Sato titled "Goal Recognition from Incomplete Action Sequences by Probabilistic Grammars" was the runner-up for the best student paper.

We would like to thank the local organizers in Nancy, Amadeo Napoli, Chedy Raissi, and their team. They did an outstanding job with the local arrangements, and the conference would not have been possible without their hard work. We would like to thank Toon Calders, Floriana Esposito, Eyke Hllermeier, and Rosa Meo, who were the program co-chairs of ECML/PKDD 2014, for agreeing to co-locate the two conferences. The two conferences had not been co-located in over a decade, and the complementary nature of the conferences led to fruitful interactions among conference participants. We must thank the members of the Technical Program Committee for providing high-quality and timely reviews. Finally, we would like to thank Jan Van Haaren for handling the paper submission process and Constantin Comendant for maintaining the conference web site.

May 2015

Jesse Davis
Jan Ramon

Organization

Program Chairs

Jesse Davis   KU Leuven, Belgium
Jan Ramon   KU Leuven, Belgium

Local Chairs

Amedeo Napoli   Inria Nancy Grand Est/LORIA, France
Chedy Rassi   Inria Nancy Grand Est/LORIA, France

Local Organizing Committee

Anne-Lise Charbonnier   Inria Nancy Grand Est, France
Louisa Touioui   Inria Nancy Grand Est, France

Program Committee

Erick Alphonse   LIPN - UMR CNRS 7030, France
Annalisa Appice   University Aldo Moro of Bari, Italy
Hendrik Blockeel   KU Leuven, Belgium
Rui Camacho   LIACC/FEUP University of Porto, Portugal
James Cussens   University of York, UK
Luc De Raedt   KU Leuven, Belgium
Inês Dutra   CRACS INES-TEC LA and Universidade do Porto, Portugal
Saso Dzeroski   Jozef Stefan Institute, Slovenia
Nicola Fanizzi   Università di Bari, Italy
Stefano Ferilli   Università di Bari, Italy
Peter Flach   University of Bristol, UK
Nuno Fonseca   EMBL-EBI, UK
Tamas Horvath   University of Bonn and Fraunhofer IAIS, Germany
Katsumi Inoue   National Institute of Informatics, Japan
Nobuhiro Inuzuka   Nagoya Institute of Technology, Japan
Andreas Karwath   University of Mainz, Germany
Kristian Kersting   Fraunhofer IAIS and Technical University of Dortmund, Germany
Ross King   University of Manchester, UK
Ekaterina Komendantskaya   University of Dundee, UK
Stefan Kramer   University of Mainz, Germany
Ondřej Kuželka   KU Leuven, Belgium

Nada Lavrač	Jozef Stefan Institute, Slovenia
Francesca Alessandra Lisi	Università degli Studi di Bari, Italy
Donato Malerba	Università degli Studi di Bari, Italy
Stephen Muggleton	Imperial College London, UK
Sriraam Natarajan	Wake Forest University School of Medicine, USA
Aline Paes	Universidade Federal Fluminense, Brazil
Bernhard Pfahringer	University of Waikato, New Zealand
Ganesh Ramakrishnan	IIT Bombay, India
Oliver Ray	University of Bristol, UK
Fabrizio Riguzzi	University of Ferrara, Italy
Celine Rouveirol	LIPN, Université Paris 13, France
Alessandra Russo	Imperial College London, UK
Chiaki Sakama	Wakayama University, Japan
Vítor Santos Costa	Universidade do Porto, Portugal
Jude Shavlik	University of Wisconsin Madison, USA
Takayoshi Shoudai	Kyushu University, Japan
Ashwin Srinivasan	Indraprastha Institute of Information Technology, India
Alireza Tamaddoni-Nezhad	Imperial College London, UK
Tomoyuki Uchida	Hiroshima City University, Japan
Guy Van den Broeck	KU Leuven, Belgium
Jan Van Haaren	KU Leuven, Belgium
Christel Vrain	LIFO - University of Orléans, France
Stefan Wrobel	Fraunhofer IAIS and University of Bonn, Germany
Akihiro Yamamoto	Kyoto University, Japan
Gerson Zaverucha	PESC/COPPE - UFRJ, Brazil
Filip Železný	Czech Technical University, Czech Republic

Web Chairs

Jan Van Haaren	KU Leuven, Belgium
Constantin Comendant	KU Leuven, Belgium

Contents

Reframing on Relational Data

Chowdhury Farhan Ahmed, Clément Charnay,
Nicolas Lachiche[(✉)], and Agnès Braud

ICube Laboratory, University of Strasbourg, Strasbourg, France
{cfahmed,charnay,nicolas.lachiche,agnes.braud}@unistra.fr

Abstract. Construction of aggregates is a crucial task to discover knowledge from relational data and hence becomes a very important research issue in relational data mining. However, in a real-life scenario, dataset shift may occur between the training and deployment environments. Therefore, adaptation of aggregates among several deployment contexts is a useful and challenging task. Unfortunately, the existing aggregate construction algorithms are not capable of tackling dataset shift. In this paper, we propose a new approach called reframing to handle dataset shift in relational data. The main objective of reframing is to build a model once and make it workable in many deployment contexts without retraining. We propose an efficient reframing algorithm to learn optimal shift parameter values using only a small amount of labelled data available in the deployment. The algorithm can deal with both simple and complex aggregates. Our experimental results demonstrate the efficiency and effectiveness of the proposed approach.

Keywords: Relational data mining · Aggregates · Dataset shift · Supervised learning · Stochastic optimization

1 Introduction

Data mining discovers hidden knowledge from databases. Relational data mining [7] is an important field of data mining which aims at extracting interesting and useful knowledge from relational databases. In a relational database, data are stored in multiple relations/tables and connected through some common keys/fields. One to many relationships are a special kind of link where each tuple of a primary table may be linked to several keys of a secondary table. An example is *Department* table and *Student* table. Here *Department* is a primary table and one department has many students in the *Student* table which implies a one to many relationship. This type of relationship is very useful for representing several real-life scenarios for example customers and purchases in market basket databases, urban blocks and buildings in geographical databases, molecules and atoms in chemical databases, phone numbers and call records in telecommunication databases, and so on.

Aggregates [4,8,12,19,20] are used to carry information from the secondary to the primary table. Consider a given customer has many transactions in the

© Springer International Publishing Switzerland 2015
J. Davis and J. Ramon (Eds.): ILP 2014, LNAI 9046, pp. 1–15, 2015.
DOI: 10.1007/978-3-319-23708-4_1

Purchase table and we want to add this information in the primary table *Customer*. Then the sum aggregate can be used to accumulate it as an attribute *total_purchase* in the primary table. Complex aggregates [4,19,20] may have multiple conditions and/or thresholds inside them. They are very useful for propositionalization [8,12,13]. Consider the following complex aggregate CA_1. It can be used to define a threshold of two classes of urban blocks in a city. For example, urban blocks satisfying CA_1 fall into class X (e.g., individual housing area), and otherwise into class Y (e.g., apartment housing area).

$$CA_1 : count(buildings\ such\ that\ area < 200) > 30$$

On the other hand, reuse of learnt knowledge is of critical importance in the majority of knowledge-intensive application areas, particularly because the operating context can be expected to vary from training to deployment. Dataset shift [14] is a crucial example of this phenomenon where training and testing datasets follow different distributions. Consider the above complex aggregate CA_1 is used to train a classifier in City 1 and we need to classify the data of City 2 and City 3 while the average area of individual houses in City 2 is comparatively smaller than City 1 and housing in a given area is more compact in City 3 than City 1. Suppose CA_2 and CA_3 represent the appropriate complex aggregates to classify the data of City 2 and City 3, respectively.

$$CA_2 : count(buildings\ such\ that\ area < 100) > 30$$
$$CA_3 : count(buildings\ such\ that\ area < 200) > 50$$

Existing relational data mining algorithms use retraining (building a new model in the deployment from the very beginning) to classify the data of City 2 or City 3. Can we use the existing model of City 1 and avoid this costly retraining? This motivates us to design a new approach of reframing on relational data. The contributions of our approach are described as follows

- We develop a new idea of reframing on relational data, and design an efficient algorithm, called Reframing Aggregates with Stochastic Hill Climbing (RASHC) for reframing both the aggregate condition and the aggregate output.
- It can handle both simple and complex aggregates.
- Both classification and regression tasks can be addressed.
- Our approach has the capability of tackling different changes in data distributions and decision functions and make the existing model workable in different deployment environments.
- It can discover optimal shift parameter values although a small amount of labelled deployment data are available.
- We experimentally show the efficiency and effectiveness of the proposed algorithm using synthetic and real-life datasets. In particular, we present the existence of dataset shift problem in a real-life dataset [1,9] and capability of our algorithm to approximate these unknown real-life shifts accurately.

The remainder of this paper is organized as follows. Section 2 discusses related work. In Sect. 3, we describe our proposed reframing approach for relational data. In Sect. 4, our experimental results are presented and analyzed. Finally, conclusions are drawn in Sect. 5.

2 Related Work

A way of mining relational databases is propositionalization [8,12,13] which transforms the data of primary and secondary tables into a single attribute value table. There are some existing methods of propositionalization such as Relaggs [12], Cardinalization and Quantiles [8]. TILDE [3] introduced a logical representation for relational decision trees and has been extended to handle complex aggregates. A method called FORF (first order random forests) [19] has been proposed for learning relational classifiers with complex aggregates. Another research work was done to refine aggregate conditions in relational learning [20]. Charnay et al. [4] proposed a method for incremental construction of complex aggregates.

On the other hand, several methods [5,11,14,18,22] have been proposed to handle the situation where training and testing environments are different. Input variable shift is most often known as covariate shift [14] in the data mining and machine learning literature. Some research works have been done to tackle covariate shift [2,14,18] where shifts in input variables are considered over different deployment contexts. For example, a new variant of cross validation, called Importance Weighted Cross Validation (IWCV), is proposed by assuming that the ratio of the test and training input densities is known at training time [18]. According to this ratio, it applies different weights to each input during cross validation to handle covariate shift. Another method, called Integrated Optimization Problem (IOP), finds contribution of each training instance to the optimization problem ideally needs to be weighted with its test-to-training density ratio [2]. It uses a discriminative model that characterizes the possibility of an instance to occur in the test set rather than in the training set. Instantiating the general optimization problem leads to a kernel logistic regression and an exponential model classifier for covariate shift [2]. In order to cope with covariate shift, Kernel Mean Matching (KMM) approach [10] reweighs the training data such that the means of the training and test data in a reproducing kernel Hilbert space (i.e., in a high dimensional feature space) are close.

Recently, an input transformation based approach, called GP-RFD (Genetic Programming-based feature extraction for the Repairing of Fractures between Data), has been proposed which can handle general dataset shift [16]. At first, it creates a classifier C for a training dataset A and considers B as a test dataset with a different distribution. To discover optimal transformation, it applies several genetic operators (e.g., selection, crossover, mutation) on B and creates dataset S. Subsequently, it applies C on dataset S and calculates the accuracy in order to determine the best transformation. This approach is computationally expensive and needs a large amount of deployment data to be labelled. In addition, another related research area is transfer learning [6,17] which learns a new model for the deployment data reusing knowledge from the base model, but it often (re)trains a new model rather than adapting the original.

However, the existing dataset shift adapting algorithms need retraining to be applied in a different deployment environment and hence are not suitable for reframing which aims at building a model once and make it (i.e., the same model) workable in several deployment environments without retraining. Moreover, they are not capable of tackling relational data. Therefore, we propose a new approach called reframing to handle dataset shift on relational data.

3   Our Proposed Approach

The main reason behind changing the condition and/or output of an aggregate is the change in data distribution. Suppose M is a model built with a base learner C using some labelled training data T_{tr}; and CA_{tr} and CA_d are complex aggregates to classify the training and deployment data, respectively. In this example, we have shown one aggregate condition for simplicity. But, there may be multiple aggregate conditions in a complex aggregate. However, if we want to use M to classify the deployment data, we have to transform the shifted parameter values of CA_d to the appropriate parameter values of CA_{tr}. By using a transformation function T and a few labelled deployment data T_d, our approach will find optimal shift parameter values for this transformation. Thus, it will be able to use the existing model that has been built with the training data. In the first and second subsections, we demonstrate the concept of reframing aggregate condition and output, respectively. Finally, our proposed reframing algorithm is presented.

$$CA_{tr} : AggF(buildings \; such \; that \; attVal_{tr} < aggCond_{tr}) > aggOutCond_{tr}$$
$$CA_d : AggF(buildings \; such \; that \; attVal_d < aggCond_d) > aggOutCond_d$$

3.1   Reframing Aggregate Condition

For reframing aggregate condition, T transforms the input attribute value $attVal_d$ of deployment to the appropriate (approximate) corresponding input attribute value $attVal_{\widehat{tr}}$ of training by an equation e.g., Eq. 1 so that they can properly be classified.

$$attVal_{\widehat{tr}} = \alpha(attVal_d) + \beta \tag{1}$$

Hence, for reframing the aggregate condition, our reframing model M evaluates the following

$$AggF(T(attVal_d) < aggCond_{tr}) = AggF((\alpha(attVal_d) + \beta) < aggCond_{tr})$$
$$= AggF(attVal_{\widehat{tr}} < aggCond_{tr})$$

The main challenging task is to find optimal shift parameter values (values of α, β in this case) so that it can transform the deployment data with these values and able to predict $attVal_{\widehat{tr}}$ very close to $attVal_{tr}$. Our method can

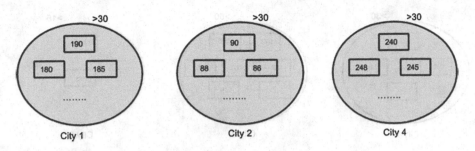

Fig. 1. Reframing aggregate condition.

perform this task properly. In addition, it can learn multiple $\{\alpha, \beta\}$ pairs if there are multiple aggregate conditions in a complex aggregate. Hence, it enables M to be built once and used over several deployment contexts without any modification. On the other hand, retraining means building a new model M_{new} for every deployment context using the few labelled data available over there while not using any knowledge of M at all. And, using a base model to other contexts means that applying model M to other deployment contexts directly, i.e., without any reframing or retraining. Accordingly, reframing will be useful to handle most of the cases of dataset shift compared to apply the base model or retraining.

$$CA_4 : count(buildings\ such\ that\ area < 250) > 30$$

Now, consider the scenario in Fig. 1 where a training environment is City 1 and deployment environments are City 2 and City 4. Complex aggregates of these cities are described in CA_1, CA_2 and CA_4, respectively. An individual housing block in City 1 contains at least 30 buildings with area less than 200. But in City 2, buildings are comparatively smaller (half compared to the building area of City 1) and in City 4 they are comparatively larger(larger than 50 units compared to the building area of City 1). However, their cardinality (aggregate output) is same like City 1 (i.e., 30). We can use the following transformation to re-scale the deployment data to fit in our existing model. For City 2, we can use $\alpha = 2$ and $\beta = 0$ (Eq. 1), so that a building area of 100 in City 2 can be mapped to its appropriate value of 200 ($100 \times 2 + 0$) in City 1 and properly be classified by the existing model of City 1. Similarly, we can use $\alpha = 1$ and $\beta = -50$ for reframing the data of City 4.

3.2   Reframing Aggregate Output

Here, we have to transform the aggregate output with T in a similar way as follows

$$\begin{aligned}
T(AggF(attVal_d < aggCond_d)) &= T(aggOut_d) \\
&= \alpha(aggOut_d) + \beta \\
&= aggOut_{\widehat{tr}}
\end{aligned} \tag{2}$$

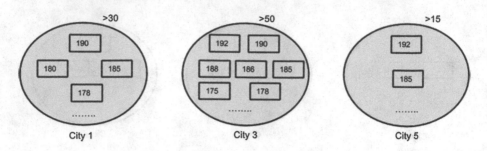

Fig. 2. Reframing aggregate output.

Our reframing approach tries to make $aggOut_{\widehat{tr}}$ very close to $aggOut_{tr}$ so that it can properly be classified by $aggOutCond_{tr}$. Consider the scenario in Fig. 2 where a training environment is City 1 and deployment environments are City 3 and City 5. Complex aggregates of these cities are described in CA_1, CA_3 and CA_5, respectively. A residential housing block in City 1 contains at least 30 buildings with area less than 200. But in City 3, even though buildings have the same size, the blocks are very crowded (too many buildings in a block) and contain at least 50 buildings. On the other hand, the reverse situation occurs in City 5, i.e., even though buildings have the same size, the blocks contain fewer buildings compared to City 1 (at least 15 buildings in a block).

$$CA_5 : count(buildings\ such\ that\ area < 200) > 15$$

For City 3, we can use $\alpha = 1$ and $\beta = -20$ (Eq. 2), so that an aggregate output of 50 in City 3 can be mapped to its appropriate value of 30 ($50 \times 1 - 20$) in City 1 and properly be classified by the existing model of City 1. Similarly, we can use $\alpha = 2$ and $\beta = 0$ for reframing the aggregate output of City 5.

3.3   RASHC: Our Proposed Algorithm

In this subsection, we present our reframing algorithm on relational data called RASHC (Reframing Aggregates with Stochastic Hill Climbing). We have considered a relational decision tree learner [4] to build the classification model M. For reframing, we consider there are n conditions in M, where each condition represents either an aggregate condition c or an aggregate output o. Consider the models of three cities given in Fig. 3 where City 1 is the training context and City 2 and City 3 are the deployment contexts. Please note that City 2 has different parameter values in the root node while leaf nodes have the same parameter values. On the other hand, City 3 has different parameter values in all the nodes (six parameters in three nodes). As described earlier, our method will learn optimal shift parameter values for these aggregates with a small amount of available labelled deployment data. In this example, it will learn total six (α, β) pairs. If there is no change, it can safely learn $\alpha = 1, \beta = 0$ values.

The RASHC algorithm is presented in Fig. 4. It uses a stochastic hill climbing search to find optimal shift parameter values. Indeed, the solution of our

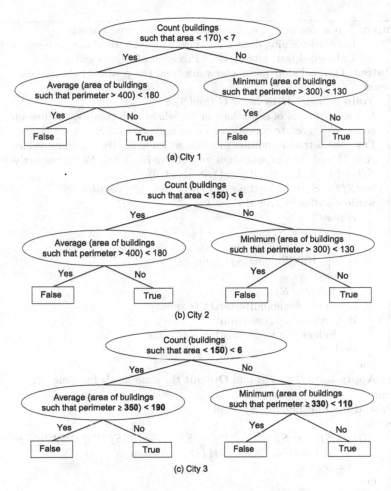

Fig. 3. Relational decision trees for three different cities.

problem space is non-convex, and according to [21] this technique provides optimal solution in most of the cases of this situation. Therefore, we have chosen
this technique for discovering our reframing solutions. The RASHC algorithm
(Fig. 4) builds a classification model M using the training labelled data (line 2).
The transformation function T uses the few available labelled data of deployment (T_d) to perform the transformation (line 4). The loop described in lines 7
to 18 learns global optimal shift parameter values. In the nested loop (lines 8 to
13), it learns a set of local optimal shift parameter value using the HillClimbing
procedure shown in lines 21 to 41. The shouldRestart procedure (invoked in line
14) updates $bestShift$ with the value of current Sft and makes the $count$ value
equal to zero, if the current Sft produces better performance than the current
$bestShift$. Otherwise, $count$ is incremented by one. It returns True if the $count$ is
less than 10, and False (i.e., $bestShift$ remains unchanged for the last 10 times)

Input: A base learner C, e.g., relational decision tree learner;
Labelled training data T_{tr};   Few labelled data of the deployment T_d;
Unlabelled test data T_{tst};   Precision of adjustment $p > 0$.

Output: Optimal shift parameter values and the predicted class labels.

1 **begin**
2   **Train** a model M by using C from T_{tr};
3   **Let** n be the no. of conditions in M; where each condition represents either an aggregate condition c or an aggregate output o;
4   **Let** T be a transformation function which shifts the attribute values $\forall c \in M$ and the output values $\forall o \in M$ by Eq. 1 and Eq. 2, respectively;
5    $Sft = (\alpha_1, \beta_1, ..., \alpha_n, \beta_n) = (1, 0, ..., 1, 0)$;
6    $bestSft = Sft$; $Continue = True$; $p = 1$; $count = 0$;
7   **while** $Continue = True$ **do**
8   **repeat**
9          $Sft_{old} = Sft$;
10   **for** $i = 1$ **to** $2n$ **do**
11   HillClimbing(Sft, i, p);
12   **end**
13   **until** $Sft = Sft_{old}$;
14       $Continue = $ shouldRestart($Sft, count$);
15   **if** $Continue = True$ **then**
16   **Select** randomly another Sft value;
17   **end**
18   **end**
19   **Apply** $bestSft$ to T_{tst} and **Output** the class labels by using M;
20 **end**
21 **Procedure** HillClimbing(Sft, i, p)
22 **begin**
23    $Sft^+ = Sft^- = Sft$; $Sft^+[i] = Sft[i] + p$; $Sft^-[i] = Sft[i] - p$;
24   **if** $Acc(M(T(Sft^+))) > Acc(M(T(Sft)))$ **then**
25       $\delta = 1$;
26   **end**
27   **else if** $Acc(M(T(Sft^-))) > Acc(M(T(Sft)))$ **then**
28       $\delta = -1$;
29   **end**
30   **else**
31   return;
32   **end**
33    $Sft_{prev} = Sft$;
34   **repeat**
35       $Incr = 1$; $Sft = Sft_{prev}$;
36   **repeat**
37          $Sft_{next} = Sft_{prev} = Sft$; $Sft[i] = Sft[i] + (Incr * p * \delta)$;
38          $Incr = Incr * 2$; $Sft_{next}[i] = Sft[i] + (Incr * p * \delta)$;
39   **until** $Acc(M(T(Sft_{next}))) < Acc(M(T(Sft)))$;
40   **until** $Incr \leq 2$;
41 **end**

Fig. 4. The RASHC algorithm.

otherwise. Accordingly, we need to start with another Sft value if the returned value is True. This part is performed in lines 15 to 17. Finally, learnt optimal parameter values of $bestShift$ are applied to transform the input values of T_{tst} and M is used to predict the class labels (line 19).

4   Experimental Results

We have performed several experiments on synthetic and real-life datasets to show the efficiency and effectiveness of our approach. We show the performance of reframing with respect to retraining and the base model. We have used a relational decision tree [4] as the base learner.

4.1   Performance in Synthetic Datasets

Three synthetic relational datasets have been generated to represent building blocks data of three different cities. The decision functions have been taken from the example given in Fig. 3. For each city, there are two tables. The primary and the secondary tables are called as $Block(\mathbf{Block\_Id}, Class)$ and $Building(\mathbf{Building\_Id}, Block\_Id, Building\_area, Building\_perimeter)$, respectively. We have generated 500 data for blocks of a particular city. For each block, the number of buildings is taken from a Geometric distribution with a mean of 10. In addition, Gaussian distributions have been used to generate the data of area ($\mu = 180$, $\sigma = 40$) and perimeter ($\mu = 450$, $\sigma = 50$) of a building.

At first, we report the accuracies of the base models of these three cities in Table 1. Now, we assume training data are available for City 1 and only a small amount of labelled data in City 2 and City 3. The number of labelled data available in the deployment is denoted as N, and we have considered $N = 10\%$ here. As mentioned in the previous sections, we have two straightforward options in this situation. Firstly, we can use the base model of City 1 directly in City 2

Table 1. Performance of the Base models of different cities.

Base	Accuracy (%)
City 1	98.4
City 2	99.8
City 3	86.4

Table 2. Performance (accuracy (%)) of the reframing algorithm.

Deployment $N = 10\%$	RASHC	Retraining	Base (City-1)
City 2	97.78	43.56	88.89
City 3	82.67	59.11	72.67

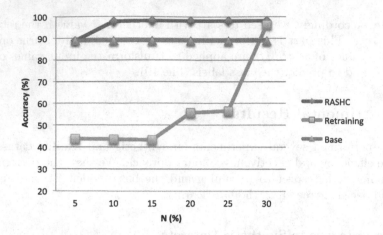

Fig. 5. Learning curves for reframing, retraining and base model (City 1) considering City 1 as training and City 2 as deployment.

and City 3. Secondly, we can develop a new model (retraining) in City 2 and City 3. The results are shown in Table 2. Please note that, the accuracy of the base model of City 1 is 98.4 %, but when it is applied in City 2 and City 3, it can achieve only 88.89 % and 72.67 % of accuracies, respectively. While accuracy of the base models of City 2 and City 3 are 99.8 % and 86.4 %, respectively. It clearly shows that dataset shift exists among these cities. Table 2 clearly indicates that the performance of retraining model with only $N = 10$ % of labelled deployment data is bad. On the other hand, our reframing algorithm RASHC achieves 97.78 % and 82.67 % of accuracies in City 2 and City 3, respectively; which are very close to the accuracy value of the base model of the corresponding cities.

Subsequently, we show learning curves in Fig. 5 for reframing, retraining and base model (City 1) considering City 1 as training and City 2 as deployment in order to represent the learning capabilities of these models more elaborately. By using the base classifier, we can only get an accuracy of 88.89 % and it is constant for all N as it does not learn anything from the labelled deployment data. From Fig. 3, we can see that optimal shift parameter values from City 2 to City 1 would be $\beta = 20$ and $\beta = 1$ for aggregate condition and aggregate output, respectively in the root node while values of the other nodes should be kept unchanged. When $N = 5$ %, RASHC is equivalent to the base model. It also illustrates that the accuracy of RASHC does not fall below the base classifier. When $N = 10$ %, it achieves around 98 % of accuracy which is very close to the original base model of City 2. On the other hand, retraining can reach nearly RASHC when $N = 30$ %.

4.2 Performance in a Real-Life Dataset with Artificial Shifts

Here, we have used a real-life relational dataset, the PKDD'99 *Financial*[1], which contains some information of a bank regarding accounts, loans, transactions etc. The bank wants to distinguish their clients as good or bad before providing them loans according to their financial history. Therefore, we have taken two tables *Loan* (as primary table) and *Transactions* (as secondary table) for our experiment. However, the *Financial* dataset is highly imbalanced (around 90 % loans are positive) between the positive and negative classes, hence, we have taken 100 loans (50 positive and 50 negative classes) and their corresponding 5,755 transactions. This dataset does not have any dataset shift. Recently, research has been done [14–16] in non-relational domains to inject different kinds of artificial shift in real-life datasets because of the unavailability of shifts inside them. Similarly, we have injected three kinds of shift (low, medium and high) in this real-life relational dataset.

The accuracy of the base model is 91 % when no shift is used. Table 3 reports how the base model is affected by the extremeness of shift values. Like the previous experiments, we have used $N = 10\%$ here. The performance of retraining shows that it is not affected by the shifts as it rebuilds the model every time with the labelled data available in the deployment rather than learning the shifts. Furthermore, it does not use any knowledge from the training section. But obviously, it cannot rebuild a good new model with this few available labelled data. On the other hand, it is noticeable that our RASHC algorithm has learnt optimal shift parameter values in all the cases and achieved good accuracies.

These experimental results demonstrate that our approach is very useful and effective to learn optimal shift parameter values, using a small amount of available labelled data (around 10 %) in deployment. It is remarkably better than retraining and its performance does not fall below the performance of the base model. Furthermore, the capability of our approach is shown for handling different changes in decision functions. Therefore, it can be built in one training context and used over several deployment contexts.

Table 3. Accuracy (%) comparison in the real-life *Financial* dataset

Shift	RASHC	Retraining	Base
Low	90	45.56	81.11
Medium	87.78	45.56	73.33
High	83.33	45.56	55.56

4.3 Performance in a Real-Life Dataset with Real Shifts

A real-life dataset, called *Bike Sharing* dataset [1,9], has been used here to show the presence of real shifts and the capability of our method to handle these shifts

---

[1] http://lisp.vse.cz/pkdd99/Challenge/chall.htm.

Table 4. Performance of Base models (different seasons) of the *Bike Sharing* dataset.

Measure	Spring	Summer	Fall	Winter
	$|Days| = 181$	$|Days| = 184$	$|Days| = 188$	$|Days| = 178$
	$|Hours| = 4242$	$|Hours| = 4409$	$|Hours| = 4496$	$|Hours| = 4232$
MAE	1068.623	1216.842	1305.509	1272.703
RMSE	1354.019	1497.588	1570.728	1568.031

properly. The *Bike Sharing* dataset contains usage logs of a bike sharing system called Capital Bike Sharing (CBS) at Washington, D.C., USA for two years (2011 and 2012). It was prepared by Fanaee-T and Gama [9], and is publicly available in UCI Machine Learning Repository [1].

The dataset contains bike rental counts hourly and daily based on the environmental and seasonal settings. The *hour.csv* file [1] contains 17,379 records where the values of bike sharing counts have been aggregated on hourly basis. Similarly the *day.csv* file [1] aggregates the bike sharing counts on daily basis and contains 731 records. The input variables contain day, hour, season, workday/holiday and some weather information such as temperature, feels like temperature, humidity and wind speed. We have considered the day format as the primary table named *Day*(**Day_Id**, *Bike rental count*) and the hour format as secondary table named *Hour*(**Hour_Id**, *Day_Id, Temperature, Feels like temperature, Humidity, Windspeed*). Our goal is to solve a regression task in this relational database which predicts the daily bike rental counts based on hourly given weather information.

Since this dataset contains data of different seasons (1:Spring, 2:Summer, 3:Fall, 4:Winter in the Season attribute), we have divided the original dataset according to different seasons in order to observe dataset shift. Table 4 shows the performance of the base models of different seasons using 10-fold cross validation. We have used the MAE (mean absolute error) and RMSE (root mean square error) measures to represent the performance in regression. To show a real-life dataset shift, we deploy the regression model trained in Spring into Summer, Fall and Winter. Please notice that the base model of Spring has a MAE value of 1068.623 in Spring (Table 4), but MAE values of 1544.563, 3193.662 and 1658.313 in Summer, Fall and Winter, respectively (Table 5). While the base models of Summer, Fall and Winter have MAE values of 1216.842, 1305.509 and 1272.703, respectively (Table 4). It clearly shows that dataset shift exists among these seasons.

Now, we run our reframing algorithm RASHC on this dataset considering Spring as training; Summer, Fall and Winter as deployments with only 10 % available labelled data at each deployment. Results are reported in Table 5. It is noticeable that the performance of RASHC is very close to the base model of that particular deployment season. For example, the base model of Summer has MAE and RMSE values of 1216.842 and 1497.588, respectively; while RASHC algorithm has MAE and RMSE values of 1314.306 and 1609.793, respectively in

Table 5. Performance comparison in the *Bike Sharing* dataset.

Deployment N = 10 %	Measure	RASHC	Retraining	Base (Spring)
Summer	MAE	1314.306	2898.77	1544.563
	RMSE	1609.793	3181.496	1954.986
Fall	MAE	1366.234	1495.57	3193.662
	RMSE	1645.177	1849.431	3691.34
Winter	MAE	1446.568	1793.186	1658.313
	RMSE	1752.089	2155.191	1988.834

this season. Moreover, reframing algorithm RASHC is significantly better than retraining and the base model of Spring.

Experimental results in this real-life dataset demonstrate that our approach is quite capable of learning optimal shift parameter values, using a small amount of labelled data (around 10 %) at deployment, in a real-life environment where the nature of a shift is unknown from source to deployment. These results also reveal the applicability of linear shift in real-life domains by clearly expressing its strength to tackle these unknown dataset shifts between one training and different deployment contexts. In addition, applicability of our approach can also be observed in sequential time series data mining.

5   Conclusions

In this paper, we have proposed a new approach of reframing on relational data. We have designed an efficient algorithm to learn optimal shift parameter values for aggregate condition and output. It can deal with both simple and complex aggregates. Our approach has the capability of tackling different changes in data distributions and decision functions, and make the existing model workable in different deployment environments. Even though a small amount of labelled data are available in the deployment, it can discover desired optimal parameter values to handle the actual dataset shift. Using synthetic and real-life datasets, we have experimentally shown the efficiency and effectiveness of the proposed algorithm. In the worst case, our approach is equivalent to the base model which is an important property for dataset shift adaptation. Furthermore, it is quite suitable for those environments where retraining is not applicable to learn the decision functions. Finally, we have presented the existence of dataset shift in a real-life dataset by considering one season as training context and other seasons as deployment contexts. We have demonstrated the capability of our approach to approximate these unknown real-life dataset shifts accurately.

Acknowledgements. This work was supported by the REFRAME project granted by the European Coordinated Research on Long-term Challenges in Information and Communication Sciences & Technologies ERA-Net (CHIST-ERA).

References

1. Bache, K., Lichman, M.: UCI machine learning repository (2013) http://archive.ics.uci.edu/ml/
2. Bickel, S., Brückner, M., Scheffer, T.: Discriminative learning under covariate shift. J. Mach. Learn. Res. **10**, 2137–2155 (2009)
3. Blockeel, H., De Raedt, L.: Top-down induction of first-order logical decision trees. Artif. Intell. **101**(1–2), 285–297 (1998)
4. Charnay, C., Lachiche, N., Braud, A.: Incremental construction of complex aggregates: counting over a secondary table. In: Late Breaking Papers of the 23rd International Conference on Inductive Logic Programming (ILP), pp. 1–6 (2013)
5. Charnay, C., Lachiche, N., Braud, A.: Pairwise optimization of bayesian classifiers for multi-class cost-sensitive learning. In: Proceedings of the 25th IEEE International Conference on Tools with Artificial Intelligence (ICTAI), pp. 499–505 (2013)
6. Davis, J., Domingos, P.: Deep transfer via second-order markov logic. In: Proceedings of the 26th Annual International Conference on Machine Learning (ICML), pp. 217–224 (2009)
7. Dzeroski, S.: Relational data mining. In: Maimon, O., Rokach, L. (eds.) Data Mining and Knowledge Discovery Handbook, pp. 887–911. Springer, New York (2010)
8. El Jelali, S., Braud, A., Lachiche, N.: Propositionalisation of continuous attributes beyond simple aggregation. In: Riguzzi, F., Železný, F. (eds.) ILP 2012. LNCS, vol. 7842, pp. 32–44. Springer, Heidelberg (2013)
9. Fanaee-T, H., Gama, J.: Event labeling combining ensemble detectors and background knowledge. Prog. Artif. Intell. **2**(2–3), 113–127 (2014)
10. Gretton, A., Smola, A., Huang, J., Schmittfull, M., Borgwardt, K., Schölkopf, B.: Covariate shift by kernel mean matching. In: Dataset Shift in Machine Learning, pp. 131–160. MIT Press, Cambridge (2009)
11. Hernández-Orallo, J.: ROC curves for regression. Pattern Recogn. **46**(12), 3395–3411 (2013)
12. Krogel, M.-A., Wrobel, S.: Transformation-based learning using multirelational aggregation. In: Rouveirol, C., Sebag, M. (eds.) ILP 2001. LNCS (LNAI), vol. 2157, p. 142. Springer, Heidelberg (2001)
13. Lachiche, N.: Propositionalization. In: Sammut, C., Webb, G.I. (eds.) Encyclopedia of Machine Learning, pp. 812–817. Springer, New York (2010)
14. Moreno-Torres, J.G., Raeder, T., Alaíz-Rodríguez, R., Chawla, N.V., Herrera, F.: A unifying view on dataset shift in classification. Pattern Recogn. **45**(1), 521–530 (2012)
15. Moreno-Torres, J.G.: Dataset shift in classification: terminology, benchmarks and methods. Ph.D thesis (2013)
16. Moreno-Torres, J.G., Llorà, X., Goldberg, D.E., Bhargava, R.: Repairing fractures between data using genetic programming-based feature extraction: a case study in cancer diagnosis. Inf. Sci. **222**, 805–823 (2013)
17. Pan, S.J., Yang, Q.: A survey on transfer learning. IEEE Trans. Knowl. Data Eng. **22**(10), 1345–1359 (2010)
18. Sugiyama, M., Krauledat, M., Müller, K.R.: Covariate shift adaptation by importance weighted cross validation. J. Mach. Lear. Res. **8**, 985–1005 (2007)
19. Van Assche, A., Vens, C., Blockeel, H., Dzeroski, S.: First order random forests: learning relational classifiers with complex aggregates. Mach. Learn. **64**(1–3), 149–182 (2006)

20. Vens, C., Ramon, J., Blockeel, H.: Refining aggregate conditions in relational learning. In: Fürnkranz, J., Scheffer, T., Spiliopoulou, M. (eds.) PKDD 2006. LNCS (LNAI), vol. 4213, pp. 383–394. Springer, Heidelberg (2006)
21. Weise, T.: Global optimization algorithms -theory and application, Second Edition (2009)
22. Zhao, H., Sinha, A.P., Bansal, G.: An extended tuning method for cost-sensitive regression and forecasting. Decis. Support Syst. 51(3), 372–383 (2011)

Inductive Learning Using Constraint-Driven Bias

Duangtida Athakravi[1](✉), Dalal Alrajeh[1], Krysia Broda[1],
Alessandra Russo[1], and Ken Satoh[2]

[1] Imperial College London, London, UK
{duangtida.athakravi07,dalal.alrajeh,k.broda,a.russo}@ic.ac.uk
[2] National Institute of Informatics, Tokyo, Japan
ksatoh@nii.ac.jp

Abstract. Heuristics such as the Occam Razor's principle have played
a significant role in reducing the search for solutions of a learning task,
by giving preference to most compressed hypotheses. For some application domains, however, these heuristics may become too weak and lead
to solutions that are irrelevant or inapplicable. This is particularly the
case when hypotheses ought to conform, within the scope of a given
language bias, to precise domain-dependent structures. In this paper
we introduce a notion of inductive learning through constraint-driven
bias that addresses this problem. Specifically, we propose a notion of
learning task in which the hypothesis space, induced by its mode declaration, is further constrained by domain-specific denials, and *acceptable
hypotheses* are (brave inductive) solutions that conform with the given
domain-specific constraints. We provide an implementation of this new
learning task by extending the ASPAL learning approach and leveraging
on its meta-level representation of hypothesis space to compute acceptable hypotheses. We demonstrate the usefulness of this new notion of
learning by applying it to two class of problems - automated revision of
software system goals models and learning of stratified normal programs.

Keywords: Inductive logic programming · Answer set programming ·
Meta-level constraints

1 Introduction

Heuristics such as language bias and minimality have widely been used for con-
trolling the search of hypotheses in a given inductive learning task. One of these
is the Occam Razor's principle, for which preferred solutions among competing
alternatives are generally those hypotheses with the fewest number of assumptions, also referred to as most compressed hypotheses. However, succinctness
alone may not guarantee that the chosen hypothesis would be *acceptable* by the

This research is partially funded by the 7th Framework EU-FET project 600792
"ALLOW Ensembles", and the EPSRC project EP/K033522/1 "Privacy Dynamics".

© Springer International Publishing Switzerland 2015
J. Davis and J. Ramon (Eds.): ILP 2014, LNAI 9046, pp. 16–32, 2015.
DOI: 10.1007/978-3-319-23708-4_2

user. The user might have specific domain-dependent criteria for selecting or rejecting a hypothesis, in addition to the hypothesis having to be part of a language bias and covering given examples. Consider the following non-monotonic learning task:

$$
B = \left\{
\begin{array}{l}
even(0) \\
num(0) \\
num(s(0)) \\
num(s(s(0))) \\
num(s(s(s(0)))) \\
num(s(s(s(s(0))))) \\
num(s(s(s(s(s(0)))))) \\
succ(X, s(X)) \leftarrow \\
\quad num(X), num(s(X))
\end{array}
\right.
\quad
E = \left\{
\begin{array}{l}
even(s(s(0))) \\
odd(s(s(s(0)))) \\
odd(s(s(s(s(s(0)))))) \\
not\ even(s(0)) \\
not\ even(s(s(s(0)))) \\
not\ odd(0) \\
not\ odd(s(s(s(s(0)))))
\end{array}
\right.
\quad
M = \left\{
\begin{array}{l}
modeh(even(+num)) \\
modeh(odd(+num)) \\
modeb(even(+num)) \\
modeb(not\ even(+num)) \\
modeb(not\ odd(+num)) \\
modeb(succ(-num, +num))
\end{array}
\right.
$$

Under brave induction [24], a most compressed hypothesis could be the set of normal clauses

$$H_1 = \{even(X) \leftarrow not\ odd(X); odd(X) \leftarrow not\ even(X)\}$$

because together with B it has a stable model which includes the examples E. Another solution of this learning task, given for instance by the hypothesis

$$H_2 = \{even(X) \leftarrow succ(Y, X), not\ even(Y); odd(X) \leftarrow succ(Y, X), even(Y)\}$$

might be more reasonable in this application domain, but it would not be preferred in the search for most compressed hypothesis as it is bigger in size than H_1. In this paper we show how domain-specific constraints on the hypothesis space can provide additional information to the learner and guide the search towards more targeted solutions. We introduce a notion of inductive learning through *constraint-driven bias*. Specifically, we define a learning task in which domain-specific constraints on the structure of the hypothesis can be expressed by denials over the search space induced by the language bias, and solutions are *acceptable hypotheses* that conform to the given domain-specific constraints. For instance, in the above example, if a constraint were specified that requires hypotheses to contain the body literal *succ*, then H_1 would be considered to be a *non-acceptable* solution. We provide an implementation of this new learning task by extending the ASPAL learning approach [10]. We leverage on the fact that constraint satisfaction is already an integral part of the abductive search [18] used by ASPAL and show that the ASPAL meta-level representation of hypothesis space is well suited for the computation of acceptable hypotheses. The mechanism of our approach is an adaptation of abductive search with constraints [17] for the purpose of inductive learning using syntactic meta-level information. We express constraints in denial form only, unlike those in [7]. The form of abduction we use is more limited than [19] as we only assume grounded abducibles. We therefore do not require constructive negation [6] and dynamic constraints. We discuss soundness and completeness results of our learning task. We illustrate how this new notion of learning through constraint-driven bias can be used to learn stratified normal programs, and demonstrate its usefulness in addressing the practical open problem of goal model refinement in requirements engineering.

2   Background

We start by briefly describing the terminology used throughout this paper. A term is a constant, a variable, or an n-ary function defined over terms, and an atom is an n-ary predicate defined over terms. A *literal* is an atom l or its default negation *not l*. A *rule* is a normal clause of the form $h \leftarrow b_1, \ldots, b_n$, where atom h is its head and literals $b_1, \ldots, b_n$ are its bodies. A *fact* is a rule with no body.

A solution to a learning task is a set of rules called a *hypothesis*. Different learning task have been presented in the literature. In this paper we focus on the meta-level learning through abductive search approach proposed in [9], and its ASP implementation ASPAL [10]. ASPAL is an abductive learning framework for nonmonotonic ILP that translates an ILP task into an equivalent abductive logic programming task and computes inductive solutions as abductive assumptions over a declarative representation of the hypothesis space (i.e. top theory $\top$). In this learning framework, an ILP task $\langle B, M, E \rangle$ consists of a set of positive and negative examples E, expressed as ground literals, a normal logic program background theory B, consisting of a set of normal clauses, and mode declarations M. Mode declarations are of the form $modeh(s)$ or $modeb(s)$ corresponding respectively to head and body mode declarations where s is a schema for a predicate containing place-holders for its arguments ('+', '−' and '#' denoting respectively input variable, output variable and constant). A literal is compatible with a mode declaration if it is an instantiation of its schema with arguments instantiated according to the place-holders. A rule is compatible with M if (i) its head literal is compatible with (i.e. its predicate appear in) a head mode declaration from M; (ii) each body literal is compatible with a body mode declaration from M; and all of its variables are correctly linked as denoted by the input and output variable place-holders in the corresponding mode declarations. An hypothesis H is a solution to the ILP task if (i) all rules in H are compatible with M; (ii) $B \cup H$ is consistent; and (iii) $B \cup H \vDash \bigwedge_{e \in E} e$[1]. ASPAL computes this learning task by an equivalent abductive task. An abductive task (ALP) is a tuple $\langle g, P, A, I \rangle$ with a set of ground facts g, called *observations*, a normal logic program P, a set of ground facts A, called *abducibles*, and a set of integrity constraints I. An abductive solution is a set $\Delta \subseteq A$ such that $P \cup \Delta$ is consistent, $P \cup \Delta \vDash g$ and $P \cup \Delta \vDash I$.

The hypothesis space R_M of the ILP task, containing all the rules compatible with M, is encoded into a *top-theory* $\top$. For every rule $r = h \leftarrow \bar{b}$ in R_M, a corresponding declarative representation $h \leftarrow \bar{b}, rule(id(h \leftarrow \bar{b}), \overline{C})$, where $\overline{C}$ is the list of constant arguments in r, is included in $\top$. In this representation, the term $id(h \leftarrow \bar{b})$ is a *rule identifier* of the form $(m_h, m_1, l_{1,1}, \ldots, l_{1,p_1}, \ldots, m_n, l_{n,1}, \ldots, l_{n,p_n})$, where m_h is the label of the mode declaration for h, m_i is the label of the mode declaration for body literal b_i in $\bar{b}$, and for each m_j, the elements $l_{j,1}, \ldots, l_{j,p_j}$ are numbers denoting the variable arguments in r linked to b_i, counting from left to right. For instance, the rules in hypothesis H_2 are encoded as $rule((even, succ, 1, not\_even, 2), e)$ and $rule((odd, succ, 1, even, 2), e)$, where e

---

[1] Throughout this paper we use $\vDash$ to refer to brave consequence.

represents the empty set of constant arguments and the mode declaration labels are *even*, *succ* and *not_even*. Given an ILP task $\langle B, M, E \rangle$, the equivalent abductive task generated by ASPAL is $\langle B', A^\top, I \rangle$ where $B' = B \cup \top \cup \{examples \leftarrow \bigwedge_{e \in E} e\}$, $A^\top$ is the set of abducible atoms $rule(id(h \leftarrow \overline{b}), \overline{C})$, I is the singleton set of integrity constraints $\{\perp \leftarrow not\ examples\}$. If $\Delta \subseteq A^\top$ is an abductive solution then the inverse translation of all abducibles in Δ, denoted $hyp(\Delta)$, is the corresponding inductive hypothesis H (see [8]).

In our constraint-driven learning approach we consider labels of literals compatible with mode declarations instead of rule identifiers, to allow us to express syntactic constraints directly of the literals that would form acceptable hypotheses. The label of a literal compatible with a mode declaration is composed of its predicate name and constant arguments. For a negated literal the label is prefixed with "*not_*". For instance, given a mode body declaration $modeb(not\ p(+int, \#int, -int))$ a compatible literal of the form $not\ p(X, 2, Z)$ will have the label $not\_p(2)$. In general, labels will be unground in the top theory (e.g. $not\_p(X)$), and their variables will be instantiated according to the compatible hypothesis that are abduced during the computation.

In what follows, for a given compatible rule r, we denote with $head(r)$ the label of its head literal and with $body(r)$ the list of labels of its body literals. In writing constraints over hypothesis structure we will often make use of variables of type labels and we will denote them with capital letters L_h and L_b. We will use the shorthand $L_h = head(r)$ and $L_b \in body(r)$, to denote that for a given compatible rule r there exists an instantiation Θ of the variables L_h and L_b such that $L_h \Theta = head(r)$ and $L_b \Theta =$ is the label of a body literal in r.

3   Learning Using Constraint-Driven Bias

Domain-dependent constraints are constraints on the hypothesis space that collectively characterise the set of *acceptable* hypotheses. We propose a set $\mathcal{L_C}$ of eight primitives for expressing domain-specific constraints. The arguments of a primitive can be either variables or constants. Variables can be head label variables (L_h), which stand for labels of an head literal in a compatible rule, or body label variables (L_{b_i}) that stand for labels of a body literal in a compatible rule. Constants are either labels referring to a specific literal or integers. The intuitive meaning of these primitives is as follows: (i) $in\_some\_rule(L_h, L_b)$ – the hypothesis must include a rule with a combination of head and body literals labelled L_h and L_b; (ii) $in\_all\_rule(L_h, L_b)$ – all rules in the hypothesis with head literal labelled L_h must have a body literal labelled L_b; (iii) $in\_same\_rule(L_h, L_{b_1}, L_{b_2})$ – in the hypothesis, body literals labelled L_{b_1} and L_{b_2} must appear together in the body of a rule defining a head literal labelled L_h; (iv) $in\_diff\_rule(L_{h_1}, L_{b_1}, L_{h_2}, L_{b_2})$ – in the hypothesis, body literals labelled L_{b_1} and L_{b_2} must appear in two different rules, with heads labelled L_{h_1} and L_{h_2} respectively; (v) $max\_body(L_h, x)$ (resp. $min\_body(L_h, x)$) – in the hypothesis rules with head literal labelled L_h have at most (resp. at least) x body literals; (vii) $max\_head(L_h, x)$ (resp. $min\_head(L_h, x)$) – the hypothesis must include at most (resp. at least) x rules with head labelled L_h. Using the primitives in $\mathcal{L}_C$ and given a background theory of an original inductive task, domain-dependent constraints are of the following form.

Definition 1. *Let B be a background knowledge expressed in a language $\mathcal{L}_B$ and M a set of mode declarations with associated constraint primitives $\mathcal{L}_C$. A domain-dependent constraint in $\mathcal{L}_C$ is $\bot \leftarrow C, \overline{C_B}$ where C is a primitive of $\mathcal{L}_C$ or its negated form and $\overline{C_B}$ is a conjunction of literals from the language $\mathcal{L}_B$.*

The following definition states what we mean by an hypothesis (i.e. set of clauses), compatible with a given mode declaration M, to satisfy a domain-dependent constraint.

Definition 2. *Let B be a background knowledge, H an hypothesis (set of normal clauses) and $\bot \leftarrow C, \overline{C_B}$ a domain-dependent constraint ic. The hypothesis H satisfies ic if and only if either $B \cup H \not\models \overline{C_B}$ or one of the following cases holds:*

- $C = not\ in\_some\_rule(L_h, L_b)$ *and there exists $r \in H$ s.t. $L_h = head(r)$ and $L_b \in body(r)$;*
- $C = not\ in\_all\_rule(L_h, L_b)$ *and either there is no rule $r \in H$ s.t. $L_h = head(r)$ or for all $r \in H$ s.t. $L_h = head(r)$ it is the case that $L_b \in body(r)$;*
- $C = not\ in\_same\_rule(L_h, L_{b_1}, L_{b_2})$ *and either there are no $r \in H$ with $L_h = head(r)$ with $L_{b_1} \in body(r)$ or $L_{b_2} \in body(r)$, or for all $r \in H$ s.t. $L_h = head(r)$ if $L_{b_1} \in body(r)$ then $L_{b_2} \in body(r)$, and vice versa;*
- $C = not\ in\_diff\_rule(L_{h_1}, L_{b_1}, L_{h_2}, L_{b_2})$ *and for some $r_1 \in H$ s.t. $L_{h_1} = head(r_1)$ and $L_{b_1} \in body(r_1)$ (resp. $L_{h_2} = head(r_2)$ and $L_{b_2} \in body(r_1)$) then it must also be the case that for some different rule $r_2 \in H$ with $L_{h_2} = head(r_2)$ (resp. $L_{h_1} = head(r_2)$) and $L_{b_2} \in body(r_2)$ (resp. $L_{b_1} \in body(r_2)$);*
- $C = not\ max\_body(L_h, x)$ *(resp. $not\ min\_body(L_h, x)$) and for all $r \in H$ if $L_h = head(r)$ then $|body(r)| \leq x$ (resp. $|body(r)| \geq x$);*
- $C = not\ max\_head(L_h, x)$ *(resp. $not\ min\_head(L_h, x)$) and for $R = \{r | r \in H$ and $L_h = head(r)\}$ then $|R| \leq x$ (resp. $|R| \geq x$).*

Primitives of $\mathcal{L}_C$ can also occur positively in a domain-dependent constraint. If ic is $\bot \leftarrow C$, where C is one of the primitives, then a hypothesis H satisfies ic if H does not satisfy $\bot \leftarrow not\ C$. For instance, if C is the primitive $in\_some\_rule(L_h, L_b)$, then H satisfies $\bot \leftarrow in\_some\_rule(L_h, L_b)$ if H does not satisfy $\bot \leftarrow not\ in\_some\_rule(L_h, L_b)$. That is, there is not a rule $r \in H$ where $L_h = head(r)$ and $L_b \in body(r)$. We can now define the notion of *constraint biased search space*. Intuitively, this is the set of rules, compatible with a given mode declaration, that satisfies the given domain-dependent constraints.

Definition 3. *Let $\mathcal{P}(R_M)$ be the power set of rules compatible with M and ic be a domain-dependent constraint. The constraint biased search space with respect to ic, denoted $[\mathcal{P}(R_M)]_{ic}$, is the set $S \subseteq \mathcal{P}(R_M)$ where each H in S satisfies ic.*

This can be extended to the notion of constraint biased search space for a set of integrity constraints.

Definition 4. *The constraint biased search space of a set IC of domain-dependent constraints, denoted $[\mathcal{P}(R_M)]_{IC}$, is the intersection of the constraint biased search spaces with respect to each domain-dependent constraint in IC:*

$$[\mathcal{P}(R_M)]_{IC} = \bigcap_{ic \in IC} [\mathcal{P}(R_M)]_{ic}$$

We can now define our notion of learning task with constraint-driven bias and acceptable solutions.

Definition 5. *A learning task with constraint-driven bias is a tuple $\langle B, M, E, IC \rangle$ where B is background knowledge, M is a set of mode declarations, E is a set of examples and IC is a set of domain-dependent constraints. H is an acceptable solution if and only if (i) $B \cup H \models \bigwedge_{e \in E} e$; and (ii) $H \in [\mathcal{P}(R_M)]_{IC}$.*

Consider the example in Sect. 1. The constraint $\perp \leftarrow not\ in\_all\_rule$ $(L_h, succ)$ will rule out the hypothesis $H_1 = \{even(X) \leftarrow not\ odd(X); odd(X) \leftarrow not\ even(X)\}$.

4   Implementation

In this section we describe how the ASPAL system has been extended to compute acceptable solutions of a learning task with constraint-driven bias. In particular, we show how the primitives described in Sect. 3 can naturally be encoded as meta-constraints on the declarative representation of the hypothesis space used in ASPAL. Table 1 gives the declarative semantics, expressed in ASP, of each primitive in $\mathcal{L}_C$. This declarative encoding makes use of the following meta-predicates: $is\_rule(R, L_h)$ represents a clause with identifier R and head literal labelled L_h; $in\_rule(R, L_h, L_b, I)$ represents a clause with identifier R, head literal labelled L_h, and body literal labelled L_b at position I. Note that the unique identifier R and the body literal position I are automatically generated by the learning system for each clause (and body) in the hypothesis space R_M. Given a rule $r = h \leftarrow b_1, \ldots, b_n$, with identifier R_{id}, of a learned hypothesis H, the associated meta-level information is the set of (ground) instances $\{in\_rule(R_{id}, L_h, L_{b_1}, 1), \ldots, in\_rule(R_{id}, L_h, L_{b_n}, n), is\_rule(R_{id}, L_h)\}$. The meta-level information for a hypothesis H, denoted with $meta(H)$, is the union of the meta-level information of its rules.

The top theory $\top_M$ is given by $\top \cup \mathcal{M}$ where $\top$ is the ASPAL top-theory as described in Sect. 2 and $\mathcal{M}$ includes for each rule $h \leftarrow b_1, \ldots, b_n, rule(id(h \leftarrow b_1, \ldots, b_n), \bar{C})$ in $\top$, the following rule, which relates a rule in R_M with its meta-level information.

$$n + 1\{is\_rule(i(\bar{C}), l_h), in\_rule(i(\bar{C}), l_h(\bar{C}_h), l_{b_1}(\bar{C}_{b_1}), 1),$$
$$\ldots, in\_rule(i(\bar{C}), l_h(\bar{C}_n), l_{b_n}(\bar{C}_{b_n}), n)\}n + 1 \leftarrow rule(id(h \leftarrow b_1, \ldots, b_n), \bar{C})$$

During the computation of a hypothesis, if the encoding $rule(id(h \leftarrow b_1, \ldots, b_n)$ of a rule r is abduced, then the above rule in $\mathcal{M}$ will enforce all the associated

Table 1. Declarative semantics of primitives of $\mathcal{L}_C$

Primitive predicate $\leftarrow P, \overline{C_B}$	Declarative semantics of $\leftarrow P, \overline{C_B}$
$\bot \leftarrow not\ in\_some\_rule(L_h, L_b), \overline{C_B}$	$ic\_cond \leftarrow in\_rule(Rid, L_h, L_b, I)$
	$\bot \leftarrow not\ ic\_cond, \overline{C_B}$
$\leftarrow in\_some\_rule(L_h, L_b), \overline{C_B}$	$\leftarrow in\_rule(Rid, L_h, L_b, I), \overline{C_B}$
$\bot \leftarrow not\ in\_all\_rule(L_h, L_b), \overline{C_B}$	$\bot \leftarrow is\_rule(Rid, L_h),$
	$\qquad not\ in\_rule(Rid, L_h, L_b, I), \overline{C_B}$
$\bot \leftarrow in\_all\_rule(L_h, L_b), \overline{C_B}$	$ic\_cond \leftarrow is\_rule(Rid, L_h)$
	$ic\_cond \leftarrow is\_rule(Rid, L_h),$
	$\qquad\qquad not\ in\_rule(Rid, L_h, L_b, I)$
	$\bot \leftarrow not\ ic\_cond, \overline{C_B}$
$\bot \leftarrow not\ in\_same\_rule(L_h, L_{b_1}, L_{b_2}), \overline{C_B}$	$\bot \leftarrow in\_rule(Rid, L_h, L_{b_1}, I_1),$
	$\qquad not\ in\_rule(Rid, L_h, L_{b_2}, I_2), \overline{C_B}$
	$\bot \leftarrow not\ in\_rule(Rid, L_h, L_{b_1}, I_1),$
	$\qquad in\_rule(Rid, L_h, L_{b_2}, I_2), \overline{C_B}$
$\bot \leftarrow in\_same\_rule(L_h, L_{b_1}, L_{b_2}), \overline{C_B}$	$\bot \leftarrow in\_rule(Rid, L_h, L_{b_1}, I_1),$
	$\qquad in\_rule(Rid, L_h, L_{b_2}, I_2), \overline{C_B}$
$\bot \leftarrow not\ in\_diff\_rule(L_{h_1}, L_{b_1}, L_{h_2}, L_{b_2}), \overline{C_B}$	$ic\_cond \leftarrow in\_rule(Rid_1, L_{h_1}, L_{b_1}, I_1),$
	$\qquad\qquad in\_rule(Rid_2, L_{h_2}, L_{b_2}, I_2),$
	$\qquad\qquad Rid_1 \neq Rid_2$
	$\bot \leftarrow not\ ic\_cond, \overline{C_B}$
$\bot \leftarrow in\_diff\_rule(L_{h_1}, L_{b_1}, L_{h_2}, L_{b_2}), \overline{C_B}$	$\bot \leftarrow in\_rule(Rid_1, L_{h_1}, L_{b_1}, I_1),$
	$\qquad in\_rule(Rid_2, L_{h_2}, L_{b_2}, I_2),$
	$\qquad Rid_1 \neq Rid_2, \overline{C_B}$
$\bot \leftarrow not\ max\_body(L_h, x), \overline{C_B}$	$\bot \leftarrow is\_rule(Rid, L_h),$
	$\qquad x + 1\{in\_rule(Rid, L_h, L_b, I)\}, \overline{C_B}$
$\bot \leftarrow max\_body(L_h, x), \overline{C_B}$	$ic\_cond \leftarrow is\_rule(Rid, L_h),$
	$\qquad\qquad x + 1\{in\_rule(Rid, L_h, L_b, I)\}$
	$\bot \leftarrow not\ ic\_cond, \overline{C_B}$
$\bot \leftarrow not\ min\_body(L_h, x), \overline{C_B}$	$\bot \leftarrow is\_rule(Rid, L_h),$
	$\qquad 0\{in\_rule(Rid, L_h, L_b, I)\}x - 1, \overline{C_B}$
$\bot \leftarrow min\_body(L_h, x), \overline{C_B}$	$ic\_cond \leftarrow is\_rule(Rid, L_h),$
	$\qquad\qquad 0\{in\_rule(Rid, L_h, L_b, I)\}x - 1$
	$\bot \leftarrow not\ ic\_cond, \overline{C_B}$
$\bot \leftarrow not\ max\_head(L_h, x), \overline{C_B}$	$\bot \leftarrow x + 1\{is\_rule(Rid, L_h)\}, \overline{C_B}$
$\bot \leftarrow max\_head(L_h, x), \overline{C_B}$	$ic\_cond \leftarrow x + 1\{is\_rule(Rid, L_h)\}$
	$\bot \leftarrow not\ ic\_cond, \overline{C_B}$
$\bot \leftarrow not\ min\_head(L_h, x), \overline{C_B}$	$\bot \leftarrow 0\{is\_rule(Rid, L_h)\}x - 1, \overline{C_B}$
$\bot \leftarrow min\_head(L_h, x), \overline{C_B}$	$ic\_cond \leftarrow 0\{is\_rule(Rid, L_h)\}x - 1$
	$\bot \leftarrow not\ ic\_cond, \overline{C_B}$

meta-level information of r to be also inferred if they satisfy the domain-specific constraints. That is, the hypothesis will satisfy the desired domain-dependent constraints. For example, the top theory for the example given in Sect. 1 will include the clauses:

$$even(X) \leftarrow succ(Y, X), not\ even(Y), rule((m1, m5, 1, m3, 2), c)$$

$$3\{is\_rule(1, even), in\_rule(1, even, succ, 1), in\_rule(1, even, not\_even, 2)\}3$$
$$\leftarrow rule((m1, m5, 1, m3, 2), c)$$

The set of abducibles in this case is $\{rule((m3, m5, 1, m6, 2), c)\}$ and the inferred meta-level information is the set $\{is\_rule(1, even), in\_rule(1, even, succ, 1), in\_rule(1, even, not\_even, 2)\}$.

Acceptable solutions of a learning task $\langle B, M, E, IC \rangle$, with constrain-driven bias IC, are computed using an *equivalent abductive task* $\langle \tilde{B}, A^\top, \tilde{I} \rangle$, where the background knowledge $\tilde{B} = B \cup \top_M \cup \{examples \leftarrow \bigwedge_{e \in E} e\}$, $A^\top$ is the same set of abducibles as in ASPAL, and $\tilde{I}$ is the set of integrity constraints $\{\perp \leftarrow not\ examples\} \cup IC^t$, where IC^t is the ASP representation of the domain-dependent constraints IC, based on the declarative semantics of the primitives given in Table 1. Acceptable hypotheses are then given by applying an inverse translation function on abductive solutions Δ of $\langle \tilde{B}, A^\top, \tilde{I} \rangle$.

In order to reason about the soundness and completeness of the implementation, we consider the constraints' ASP translation.

Proposition 1. *Let $\langle B, M, E, IC \rangle$ be a learning task with constraint-driven bias and let $\langle \tilde{B}, A^\top, \tilde{I} \rangle$ be an equivalent abductive task. For each $\Delta \subseteq A^\top$, $hyp(\Delta) \in [\mathcal{P}(R_M)]_{IC}$ if and only if $B \cup \top \cup \Delta \cup meta(hyp(\Delta)) \cup IC^t$ is satisfiable.*

Proof Outline: The proposition essentially shows that the ASP encoding of an hypothesis from a given constraint biased search space satisfies the ASP encoding of the given domain-dependent constraints. A proof outline is given which shows, by some cases, that this holds for single constraints ic with just one primitive.

- ic is $\perp \leftarrow not\ in\_some\_rule(L_h, L_b)$: $hyp(\Delta) \in [\mathcal{P}(R_M)]_{ic}$ iff, by Definitions 2 and 3, there exists $r \in hyp(\Delta)$ s.t. $L_h = head(r)$ and $L_b \in body(r)$. Therefore $in\_rule(Rid, L_h, L_b, I) \in meta(hyp(\delta))$, where $\delta \in A^\top$ and $hyp(\delta) = r$, for some value of I and Rid the rule identifier of r. Hence $\delta \cup meta(hyp(\delta)) \cup ic^t$ is satisfiable.

- ic is $\perp \leftarrow not\ in\_all\_rule(L_h, L_b)$: $hyp(\Delta) \in [\mathcal{P}(R_M)]_{ic}$ iff, by Definitions 2 and 3, either there is no rule $r \in hyp(\Delta)$ s.t. $L_h = head(r)$ or for all $r \in hyp(\Delta)$ s.t. $L_h = head(r)$ it is the case that $L_b \in body(r)$. In the first case, $is\_rule(Rid, L_h) \notin meta(hyp(\delta))$, where $\delta \in A^\top$ and $hyp(\delta) = r$, for any rule identifier Rid making $\delta \cup meta(hyp(\delta)) \cup ic$ satisfiable. In the second case, $in\_rule(Rid, L_h, L_b, I) \in meta(hyp(\delta))$, where $\delta \in A^\top$ and $hyp(\delta) = r$, for some value of I and Rid so making $\delta \cup meta(hyp(\delta)) \cup ic^t$ satisfiable.

- ic is $\perp \leftarrow not\ in\_same\_rule(L_h, L_{b_1}, L_{b_2})$: $hyp(\Delta) \subseteq [\mathcal{P}(R_M)]_{ic}$ iff, by Definitions 2 and 3, either there are no $r \in hyp(\Delta)$ with $L_h = head(r)$ with

$L_{b_1} \in body(r)$ or $L_{b_2} \in body(r)$, or for all $r \in hyp(\Delta)$ s.t. $L_h = head(r)$ if $L_{b_1} \in body(r)$ then $L_{b_2} \in body(r)$, and vice versa. In the first case either $in\_rule(Rid, L_h, L_{b_1}, I_1)) \notin meta(hyp(\delta))$, where $\delta \in A^\top$ and $hyp(\delta) = r$, for any rule identifier Rid and $in\_rule(Rid, L_h, L_{b_2}, I_2)) \notin meta(hyp(\delta))$, where $\delta \in A^\top$ and $hyp(\delta) = r$, so making $\delta \cup meta(hyp(\delta)) \cup ic^t$ is satisfiable. In the second case, both atoms $in\_rule(Rid, L_h, L_{b_1}, I_1))$ and $in\_rule(Rid, L_h, L_{b_2}, I_2))$ are in $meta(hyp(\delta))$ so making $\delta \cup meta(hyp(\delta)) \cup ic^t$ is satisfiable. The two ASP denials that form ic^t work together to ensure that both literals must be in the same rule if either one of them is.

- ic is $\bot \leftarrow not\ in\_diff\_rule(L_{h_1}, L_{b_1}, L_{h_2}, L_{b_2})$: $hyp(\Delta) \in [\mathcal{P}(R_M)]_{ic}$ iff, by Definitions 2 and 3, for some $r_1 \in hyp(\Delta)$ s.t. $L_{h_1} = head(r_1)$ and $L_{b_1} \in body(r_1)$ (resp. $L_{h_2} = head(r_2)$ and $L_{b_2} \in body(r_1)$) there must also exist some different rule $r_2 \in hyp(\Delta)$ with $L_{h_2} = head(r_2)$ (resp. $L_{h_1} = head(r_2)$) and $L_{b_2} \in body(r_2)$ (resp. $L_{b_1} \in body(r_2)$). Consider some $r_1 \in hyp(\Delta)$ s.t. $L_{h_1} = head(r_1)$ and $L_{b_1} \in body(r_1)$ (resp. $L_{h_2} = head(r_2)$). In this case, both $in\_rule(Rid, L_{h_1}, L_{b_1}, I_1)$ and $in\_rule(Rid, L_{h_1}, L_{b_2}, I_2)$ are in $meta(hyp(\delta))$, where $\delta \in A^\top$ and $hyp(\delta) = r$ for some values of Rid_1, Rid_2, I_1 and I_2, so making $\delta \cup meta(hyp(\delta)) \cup ic^t$ is satisfiable.

- ic is $\bot \leftarrow not\ max\_body(L_h, x)$ ($not\ min\_body(L_h, x)$): $hyp(\Delta) \in [\mathcal{P}(R_M)]_{ic}$ iff, by Definitions 2 and 3, for all $r \in hyp(\Delta)$ if $L_h = head(r)$ then $|body(r)| \leq x$ (resp. $|body(r)| \geq x$). Let $r \in hyp(\Delta)$ such that $L_h = head(r)$. So $is\_rule(Rid, L_h) \in meta(hyp(\delta))$, where $\delta \in A^\top$ and $hyp(\delta) = r$ and up to x $in\_rule(Rid, L_h, L_b, I) \in meta(hyp(\delta))$. Hence $\delta \cup meta(hyp(\delta)) \cup ic^t$ is satisfiable.

- ic is $\bot \leftarrow not\ max\_head(L_h, x)$ ($not\ min\_head(L_h, x)$): $hyp(\Delta) \in [\mathcal{P}(R_M)]_{ic}$ iff, by Definitions 2 and 3, $|\{r | r \in hyp(\Delta)\ \text{and}\ L_h = head(r)\}| \leq x$ (resp. $\geq x$). So, $|\{isrule(R_{id}, L_h) : isrule(R_{id}, L_h) \in meta(hyp(\delta))\}| \leq x$, where $\delta \in A^\top$ and $hyp(\delta) = r$. Hence $\delta \cup meta(hyp(\delta)) \cup ic^t$ is satisfiable.

It is easy to show that applying the above cases to each ic appearing in a given set of domain-dependent constraints, the proposition holds for the set $hyp(\Delta)$. Moreover, an integrity constraint ic^t that, in addition to a single primitive, includes background literals $\overline{C_B}$ will be satisfied either if $\overline{C_B}$ is not entailed by $B \cup hyp(\Delta)$, or if $\overline{C_B}$ is entailed by $B \cup hyp(\Delta)$ and the single primitive is satisfied. Thus $hyp(\Delta) \in [\mathcal{P}(R_M)]_{ic}$ iff $B \cup \top \cup \Delta \cup meta(hyp(\Delta)) \cup ic^t$ is satisfiable, and consequently $hyp(\Delta) \in [\mathcal{P}(R_M)]_{ic}$ iff $B \cup \top \cup \Delta \cup meta(hyp(\Delta)) \cup ic^t$ is satisfiable for all $ic \in IC$.

The next theorem shows that the ASP implementation is sound and complete with respect to our notion of acceptable solution of a learning task with constrain-driven bias.

Theorem 1. *Let $\langle B, M, E, IC \rangle$ be a learning task with constraint-driven bias. The set of rule $hyp(\Delta)$ is an acceptable solution of this task if and only if Δ is an abductive solution of the abductive task $\langle \tilde{B}, A^\top, \tilde{I} \rangle$.*

Proof Outline: We consider first the completeness case. For $hyp(\Delta)$ to be an acceptable solution, $hyp(\Delta) \in [\mathcal{P}(R_M)]_{IC}$ and $B \cup hyp(\Delta) \models \bigwedge_{e \in E} e$ bravely.

The latter means that there must exist an answer set AS of $B \cup hyp(\Delta)$ that extends $\bigwedge_{e \in E} e$. By Proposition 1 $B \cup \top \cup \Delta \cup meta(hyp(\Delta)) \cup IC^t$ is satisfiable. We also know that $B \cup \top \cup \Delta$ has an answer set that proves the examples because unfolding (answer set-preserving transformation) $\top$ with respect to Δ yields $B \cup hyp(\Delta)$. So adding $\{examples \leftarrow \bigwedge_{e \in E} e\} \cup \{\bot \leftarrow not\ examples\}$ will make $\tilde{B} \cup \Delta \cup \tilde{I}$ satisfiable, which implies that Δ is an abductive solution of $\langle \tilde{B}, A^\top, \tilde{I} \rangle$.

Consider now the soundness case. Let AS be an answer set of $\tilde{B} \cup \Delta \cup \tilde{I}$. As $\tilde{I} = \{\bot \leftarrow examples\} \cup IC^t$, and $\{examples \leftarrow \bigwedge_{e \in E} e\}$ is in $\tilde{B}$, then AS satisfies the examples. The examples are only satisfiable by $B \cup \top \cup \Delta$, and unfolding $\top$ with respect to Δ yields $B \cup hyp(\Delta)$, thus $B \cup hyp(\Delta)$ satisfies the examples. Since $\tilde{B} \cup \Delta \cup \tilde{I}$ is satisfiable then $B \cup \top \cup \Delta \cup meta(hyp(\Delta)) \cup IC^t$ is satisfiable. Hence it follows by Proposition 1 that $hyp(\Delta) \in [\mathcal{P}(R_M)]_{IC}$, and therefore according to Definition 5 $hyp(\Delta)$ is acceptable.

5 Case Studies

5.1 Requirements Engineering

In the area of requirements engineering, goal models are directed acyclic graphs for representing links between objectives the software system is expected to meet. In these models high-level goals are linked to their sub-goals through refinement links such that satisfying the conjunction of the sub-goals is a sufficient condition for satisfying its parent. A library of goal patterns (expressed in propositional temporal logic [23]) that guarantees these refinement semantics is given in [11]. An example is given in Fig. 1. Figure 1b shows a simplified goal model for a Flight Control System (FCS) from [2] that was obtained using the refinement pattern in Fig. 1a. The symbol $\square$ means always, $\bigcirc$ means at the next time point, $\rightarrow$ is for logical implication, and *MovingOnRunway*, *ThrustEnabled* and *WheelsTurning* are propositional atoms The top-level goal, labelled g, states that whenever the plane is moving on the runway, the reverse thrust shall be enabled at the next time point. The goal g has two children labelled $c1$ and $c2$ respectively. The left child $c1$ says that whenever the plane is moving on the runway, the wheels are turning, whilst the right child $c2$ says that whenever the wheels are turning, the reverse thrust is enabled next.

One problem with early goal-driven models (such as the one in Fig. 1b) is that they may be incorrect or too weak and hence need to be continually revised until a complete and correct specification is reached. Consider a scenario in

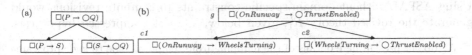

Fig. 1. (a) A goal pattern from [11]; (b) FCS goal refinement model with goals labelled g, $c1$ and $c2$.

which the runway surface is wet, the plane is moving on the runway, the wheel sensors are flagged and the wheels are not turning, hence violating the goal $c1$. In such circumstances, the idealistic or conflicting goals need to be revised to avoid any potential system failures. Not only so, but all goals dependent on the problematic goal (e.g., children, other sub-goals and parent goals) may also need to be modified to ensure that the correctness of the goal model is maintained. Very limited formal support is available for this crucial task [21]. The purpose of this case study is to demonstrate how theory revision with domain-dependent constraints on mode declarations provides formal, automated support to sound revisions of goal models, something not achievable by current non-monotonic ILP systems like ASPAL.

To revise the model, each assertion in Fig. 1b is mapped into a set of rules, using an Event Calculus formalism [20] similar to the one presented in [1], that captures the syntactic formulation of the goals[2]. The predicates $a\_holds/3$ (resp. $c\_holds/3$) represent the notion that the antecedent (resp. consequent) of the goal holds at a time point in a scenario.

$$
R = \left\{
\begin{array}{l}
a\_holds(c1, T, S) \leftarrow holds\_at(onRunway, T, S) \\
c\_holds(c1, T, S) \leftarrow holds\_at(wheelsTurning, T, S) \\
a\_holds(c2, T, S) \leftarrow holds\_at(wheelsTurning, T, S) \\
c\_holds(c2, T, S) \leftarrow holds\_at(thrustEnabled, T, S)
\end{array}
\right\}
$$

Violating scenarios are included in the background theory as a series of *happens* facts, e.g.,[3]

$$
B = \left\{
\begin{array}{ll}
happens(rain, 0, s1) & happens(landPlane, 1, s1) \\
happens(switchPulseOn, 1, s1) & happens(enableReverseThrust, 2, s1) \\
...
\end{array}
\right\}
$$

The aim of the learning task is to revise the goals appearing in Fig. 1b, by replacing the condition *WheelsTurning* in the sub-goals $c1$ and $c2$ which is not true in the violating scenario with another that does. We use our new notion of learning to compute acceptable revisions of the goal models, i.e., revisions that preserve the semantics of the goal model. An example constraint is that literals appearing in the body of a rule with head label a_holds must also appear in the body of a rule with head label c_holds. This is given below.

$$
IC = \left\{
\begin{array}{l}
\bot \leftarrow in\_diff\_rule(a\_holds(C1), BL1, c\_holds(C2), BL2), \\
\quad right\_child\_of(G, C1), left\_child\_of(G, C2), BL1 \neq BL2, C1 \neq C2 \\
\bot \leftarrow in\_diff\_rule(c\_holds(C1), BL1, a\_holds(C2), BL2), \\
\quad left\_child\_of(G, C1), right\_child\_of(G, C2), BL1 \neq BL2, C1 \neq C2 \\
\bot \leftarrow not\ min\_body(a\_holds(C), 1) \\
\bot \leftarrow not\ min\_body(c\_holds(C), 1)
\end{array}
\right\}
$$

Using ASPAL without domain-specific constraints to compute revisions would generate the revised theory R'_1, given below, as most compressed hypothesis,

---

[2] We do not present here details of the mapping from temporal logic to logic programs as this is outside the scope of this paper.

[3] The complete program is available from http://www.doc.ic.ac.uk/~da04/ilp14/goalmodel.lp.

which does not preserve the refinement semantics of the goal model (where the same assertion S must appear in both $c1$ and $c2$ according to the pattern in Fig. 1a), nor the correctness of goal models (requiring the parent goal to hold the conjunction of its children holds).

$$R'_1 = \begin{cases} a\_holds(c1, T, S) \leftarrow holds\_at(onRunway, T, S) \\ c\_holds(c1, T, S) \\ a\_holds(c2, T, S) \leftarrow holds\_at(wheelsPulseOn, T, S) \\ c\_holds(c2, T, S) \leftarrow holds\_at(thrustEnabled, T, S) \end{cases}$$

On the other hand, using the constraint-driven bias extension of ASPAL with domain-specific constraints that capture the goal-model refinement semantics we are able to compute the acceptable revised theory R'_2, given below, which although not minimal, satisfies the refinement semantics.

$$R'_2 = \begin{cases} a\_holds(c1, T, S) \leftarrow holds\_at(onRunway, T, S) \\ c\_holds(c1, T, S) \leftarrow holds\_at(wheelsPulseOn, T, S) \\ a\_holds(c2, T, S) \leftarrow holds\_at(wheelsPulseOn, T, S) \\ c\_holds(c2, T, S) \leftarrow holds\_at(thrustEnabled, T, S) \end{cases}$$

5.2 Hypothesis Stratification

In this second case study, we show more generally how constraints on mode declarations could be used to compute stratified programs [3] as acceptable solutions of a learning task. In our non-monotonic setting this is useful to guarantee the uniqueness of a hypothesis' interpretation. We consider two instances of this learning task. In the first case the given background knowledge is empty and we want to compute a set of stratified programs that proves the examples. In the second case the learning task has a given (stratified) background knowledge and we want to compute a hypothesis such that the union of the hypothesis and given background knowledge is stratified and proves the examples.

Let's consider the first case. Example of domain-dependent constraints, for guiding the search to stratified programs, are as follows:

$$\bot \leftarrow in\_some\_rule(H, B), is\_positive(B), stratum(H, L), not\ defined(B, L)$$
$$\bot \leftarrow in\_some\_rule(H, B), is\_negative(B), stratum(H, L), not\ predefined(B, L)$$
$$\bot \leftarrow not\ min\_head(H, 1), stratum(H, L), L \neq 1.$$

The first constraint checks that positively occurring body literals are defined within the same or lower stratum than the head stratum. The second constraint checks that negatively occurring body literals are defined in a lower stratum than the head stratum, and the third constraint ensures that literals exclusively used in the body (i.e. are not used in any head) are in the lowest stratum. Auxiliary predicates can be defined to help reasoning about basic concepts related to stratification. These include facts about labels of positive literals (using a predicate $is\_positive$), labels of negative literals (using a predicate $is\_negative$), and facts about labels of corresponding negative and positive pairs of literals (using the predicate $negative\_positive\_pair/2$). A notion of strata can be expressed using the ASP choice rule to ensure that each literal in the hypothesis is assigned a

Table 2. Mode declarations used in the two learning tasks for testing the stratification constraints. All tasks were run to find hypotheses with at most two clauses and maximum two body literals per clause.

Task	Head declaration	Body declaration
1	modeh(p(+arg,+arg))	modeb(q(+arg,+arg))
	modeh(q(+arg,+arg))	modeb(not q(+arg,+arg))
		modeb(p(+arg,+arg))
		modeb(not p(+arg,+arg))
2	modeh(p(+arg,+arg))	modeb(q(+arg,+arg))
	modeh(q(+arg,+arg))	modeb(not q(+arg,+arg))
	modeh(a(+arg,+arg))	modeb(p(+arg,+arg))
		modeb(not p(+arg,+arg))
		modeb(a(+arg,+arg))
		modeb(not a(+arg,+arg))

unique stratum; the predicate *level* asserts an integer limit on the number of strata allowed in a hypothesis, and the number of levels needed is at most equal to the number of predicates that can appear in the hypothesis:

$$1\{stratum(H, L) : level(L)\}1 \leftarrow is\_positive(H)$$
$$\perp \leftarrow stratum(H, L1), stratum(H, L2), L1 \neq L2$$

Other key auxiliary predicates are defined below. Predicates *defined* and *predefined* are used to check that positive and negative body literals are appropriately defined in a lower or equal stratum to where they are used:

$$defined(B, L1) \leftarrow level(L1), stratum(B, L2), L2 \leq L1$$
$$predefined(NegB, L1) \leftarrow level(L1), negative\_positive\_pair(NegB, B),$$
$$stratum(B, L2), L2 < L1$$

We have considered two learning tasks[4] of this kind, where no examples was given, the background included only the definition of the auxiliary predicates, described above, and the mode declaration were defined as shown in Table 2. Each learning task was solved for a different range of allowable values for variables of type *int*, from 1 to 5, 1 to 10, and 1 to 20.

Table 3 shows the number of unique hypotheses computed in each instance, with and without the constraints. In Table 3 are the recorded results of the case study[5]. For both tasks the time taken to find all hypotheses using the constraints is smaller than the learning task without constraints. We can conclude

---

[4] Clingo's option `--project` was used when running the ASP programs for solving the examples using the constraints to ensure that each hypothesis is output only once regardless of the number of ways it could be stratified.

[5] All tasks were run using the ASP solver Clingo 3 [13] on a 2.13 GHz laptop computer with 4 GB memory.

Table 3. Number of hypotheses learnt with and without using the constraints, and the time taken to learn all hypotheses for when the argument of the rules can be integer within the ranges 1 to 5, 1 to 10 and 1 to 15.

Task	Range for `arg`	Without constraints		With constraints	
		No. of hypotheses	Time (s)	No. of hypotheses	Time (s)
1	1–5	25126	3.588	8614	1.716
	1–10		9.282		4.820
	1–15		34.788		14.960
2	1–5	302860	219.306	158482	198.433
	1–10		1494.942		1082.694
	1–15		4640.889		2346.723

that for these experiments that domain-specific constraints help improving the efficiency of the task of learning stratified programs. More importantly, for the non-constrained tasks, the hypotheses produced are not all stratified, thus further parsing of the results is required to find only the stratified hypotheses.

The second type of problem we have considered is when we want to learn programs that together with an existing background knowledge are stratified. The stratified constraint has now to apply to the union of the given background and a computed hypothesis. To achieve this a meta-level representation of the existing background knowledge is constructed using is_rule and in_rule, adding instances of the predicates $is\_positive$, $is\_negative$ and $negative\_positive\_pair$ as necessary. This case study only deals with ungrounded rules, but the constraint is still applicable to rules with constants. Consider the even and odd example in Sect. 1. The predicates $is\_positive$ and $is\_negative$ can be extended as follows:

$$is\_positive(succ, succ) \qquad\qquad is\_negative(not\_even, not\_even)$$
$$is\_positive(even, even) \qquad\qquad is\_negative(not\_odd, not\_odd)$$
$$is\_positive(even, even(X)) \leftarrow num(X) \quad is\_positive(odd, odd)$$

The original domain-dependent constraints are in this case:

$$IC = \left\{ \begin{array}{l} \leftarrow in\_some\_rule(H, B\_label), is\_positive(B\_name, B\_label), \\ \quad stratum(H, L), not\ defined(B\_name, L) \\ \leftarrow in\_some\_rule(H, B\_label), is\_negative(B\_name, B\_label), \\ \quad stratum(H, L), not\ predefined(B\_name, L) \end{array} \right\}$$

Solving for hypotheses containing at most two rules and at most three body literals, a total of ten hypotheses can be found including the following:

$$\left\{ \begin{array}{l} odd(A) \leftarrow succ(B, A), succ(C, B), not\ even(C) \\ even(A) \leftarrow succ(B, A), succ(C, B), even(C) \end{array} \right\}$$
$$\left\{ \begin{array}{l} even(A) \leftarrow succ(B, A), succ(C, B), even(C) \\ odd(A) \leftarrow succ(B, A), even(B) \end{array} \right\}$$
$$\left\{ \begin{array}{l} even(A) \leftarrow succ(B, A), succ(C, B), even(C) \\ odd(A) \leftarrow not\ even(A) \end{array} \right\}$$

In order to find only the more succinct hypotheses, additional constraints could be used. For instance, the constraint $\perp \leftarrow not\ max\_body(odd, 2)$, which states that rules defining the concept of odd should have a maximum of two body literals, would rule out the hypothesis

$$\left\{ \begin{array}{l} odd(A) \leftarrow succ(B, A), succ(C, B), not\ even(C) \\ even(A) \leftarrow succ(B, A),\ succ(C, B), even(C) \end{array} \right\}.$$

6   Conclusion and Related Work

In this paper we have proposed a new notion of learning, where domain-dependent constraints can be specified as part of the learning task and implemented in terms of integrity constraints on the syntactic structure of compatible with the given mode declaration. *Acceptable* hypotheses are set of clauses that are compatible with the mode declaration, satisfy the domain-dependent constraints and (bravely) cover all the examples. We have shown how the meta-level representation and abductive search for inductive solutions is particularly suited to handle this type of learning tasks. This is demonstrated by considering the ASPAL learning system and extending it to support learning using constraint-driven bias. The new extended implementation has also been proved to be sound and complete with respect to our proposed new notion of learning. Finally, we have illustrated its applicability to two type of problems – goal model refinements in requirements engineering and computation of stratified programs from a given search space. Preliminary experimental results show that the computational time is not affected by the extra constraint satisfaction problem, but in contrast it helps in reducing the magnitude of the number of computed hypotheses.

The notion of meta-level top theory for reasoning about hypotheses and for expressing constraints on answer sets is related to the approach in [12] where meta-level information is used to express constraints and preferences. Somewhat related to our approach is the work in [15] where knowledge is represented as causal graphs and constraints are added by specifying impossible connections between nodes. Similarly to the work in [15], Metagol$_D$ [22] handles the problem of learning through a meta-interpreter top theories by lifting the entire learning task into a second-order representation and making use of a meta-interpreter – theory of allowable rule formats – for restricting the hypothesis space. Our work differs from these two approaches in that we generally use meta-level representation only to represent the hypothesis space and constraints, rather than the entire background knowledge and examples. This is obviously with the exception of the stratification problem presented, where the global property (.e. stratification) that the constraints try to maintain in the search for hypotheses by definition applies not only to the learned hypothesis but also to an existing background knowledge.

More related to our work is the notion of production fields for enforcing constraints on the hypothesis [14]: a tuple $\langle L, C \rangle$ could be expressed where L is a literal and C is the condition to be satisfied for L to be included in the hypothesis. As far as we are aware, in past works, the condition in a production field has mainly been used to specify conditions like length of the clause or depth of a term. Our approach is able to enforce similar constraints for body declarations using the primitives $max\_body/2$ and $min\_body/2$. Furthermore, as shown in our case studies more complex constraints across different clauses can also be imposed. Other more loosely related approaches that have made use of enhanced language bias to apply additional constraints to the hypothesis space include [4,5]. They both add a cardinality constraint to each mode declaration to specify the minimal or maximal number of times the mode declaration can be used in the hypothesis. Past work that use integrity constraints in ILP includes [16], which focuses on how to check for constraint satisfiability, and with respect to meta-level information only uses arguments' types. This differs from our work which directly reasons on the structure of the hypothesis, and allows for the constraints to be defined over relationships between different rules.

In this paper we have only considered using one primitive of $\mathcal{L}_C$ in each integrity constraint. For future work we intend to extend it to multiple primitives in the same constraint. This requires a more refined encoding in ASP in order to consider the primitives and their negated forms can be safely represented.

References

1. Alrajeh, D., Kramer, J., van Lamsweerde, A., Russo, A., Uchitel, S.: Generating obstacle conditions for requirements completeness. In: ICSE, pp. 705–715 (2012)
2. Alrajeh, D., Kramer, J., Russo, A., Uchitel, S.: Elaborating requirements using model checking and inductive learning. IEEE Trans. SE **39**(3), 361–383 (2013)
3. Apt, K.R., Blair, H.A., Walker, A.: Foundations of deductive databases and logic programming. In: Minker, J. (ed.) Towards a Theory of Declarative Knowledge, pp. 89–148. Morgan Kaufmann, Los Altos (1988)
4. Blockeel, H., Raedt, L.D.: Top-down induction of first-order logical decision trees. Artif. Intell. **101**(1–2), 285–297 (1998)
5. Bragaglia, S., Ray, O.: Nonmonotonic learning in large biological networks. In: 24th International Conference on Inductive Logic Programming (2014)
6. Chan, D.: Constructive negation based on the completed database. In: Proceedings of the Fifth International Conference and Symposium on Logic Programming, 1988, (2 Volumes), pp. 111–125 (1988)
7. Christiansen, H.: Executable specifications for hypothesis-based reasoning with prolog and constraint handling rules. J. Appl. Log. **7**(3), 341–362 (2009)
8. Corapi, D.: Nonmonotonic inductive logic programming as abductive search. Ph.D. thesis, Imperial College London (2011)
9. Corapi, D., Russo, A., Lupu, E.: Inductive logic programming as abductive search. In: Hermenegildo, M.V., Schaub, T. (eds.) ICLP (Technical Communications). LIPIcs, vol. 7, pp. 54–63 (2010)
10. Corapi, D., Russo, A., Lupu, E.: Inductive logic programming in answer set programming. In: Muggleton, S.H., Tamaddoni-Nezhad, A., Lisi, F.A. (eds.) ILP 2011. LNCS, vol. 7207, pp. 91–97. Springer, Heidelberg (2012)
11. Darimont, R., van Lamsweerde, A.: Formal refinement patterns for goal-driven requirements elaboration. In: FSE, pp. 179–190. ACM (1996)
12. Eiter, T., Faber, W., Leone, N., Pfeifer, G.: Computing preferred answer sets by meta-interpretation in answer set programming. TPLP **3**(4), 463–498 (2003)

13. Gebser, M., Kaminski, R., Kaufmann, B., Schaub, T.: Clingo = ASP + control: preliminary report. In: ICLP 2014, vol. 14(4–5) (2014)
14. Inoue, K.: Induction as consequence finding. ML **55**(2), 109–135 (2004)
15. Inoue, K., Doncescu, A., Nabeshima, H.: Completing causal networks by meta-level abduction. Mach. Learn. **91**(2), 239–277 (2013)
16. Jorge, A., Brazdil, P.: Integrity constraints in ILP using a Monte Carlo approach. In: Muggleton, S. (ed.) ILP 1996. LNCS, vol. 1314, pp. 137–151. Springer, Heidelberg (1996)
17. Kakas, A., Michael, A., Mourlas, C.: ACLP: abductive constraint logic programming (2000)
18. Kakas, A.C., Kowalski, R.A., Toni, F.: Abductive logic programming. J. Log. Comput. **2**(6), 719–770 (1992)
19. Kakas, A.C., Nuffelen, B.V., Denecker, M.: A-system: problem solving through abduction. In: Nebel, B. (ed.) Proceedings of the Seventeenth International Joint Conference on Artificial Intelligence, 2001, pp. 591–596. Morgan Kaufmann (2001)
20. Kowalski, R.A., Sergot, M.: A logic-based calculus of events. New Gener. Comput. **4**(1), 67–95 (1986)
21. van Lamsweerde, A., Letier, E.: Handling obstacles in goal-oriented requirements engineering. IEEE Trans. SE **26**(10), 978–1005 (2000)
22. Lin, D., Dechter, E., Ellis, K., Tenenbaum, J.B., Muggleton, S.: Bias reformulation for one-shot function induction. In: ECAI 2014. Frontiers in Artificial Intelligence and Applications, vol. 263, pp. 525–530 (2014)
23. Manna, Z., Pnueli, A.: The Temporal Logic of Reactive and Concurrent Systems. Springer, New York (1992)
24. Sakama, C., Inoue, K.: Brave induction: a logical framework for learning from incomplete information. Mach. Learn. **76**(1), 3–35 (2009)

Nonmonotonic Learning
in Large Biological Networks

Stefano Bragaglia and Oliver Ray[(⊠)]

Department of Computer Science, University of Bristol, Bristol BS8 1TH, UK
{stefano.bragaglia,oliver.ray}@bristol.ac.uk

Abstract. This paper introduces a new open-source implementation of
a nonmonotonic learning method called XHAIL and shows how it can be
used for abductive and inductive inference on metabolic networks that
are many times larger than could be handled by the preceding prototype. We summarise several implementation improvements that increase
its efficiency and we introduce an extended form of language bias that
further increases its usability. We investigate the system's scalability in
a case study involving real data previously collected by a *Robot Scientist*
and show how it led to the discovery of an error in a whole-organism
model of yeast metabolism.

Keywords: ILP · ALP · ASP · Metabolic networks · Completion ·
Revision

1 Introduction

The task of extending metabolic models to make them consistent with observed
data is an important application of ILP which has been tackled in projects
such as the *Robot Scientist* [6,7]. The task of revising (as opposed to merely
extending) such models is a nonmonotonic ILP problem addressed by methods
like eXtended Hybrid Abductive Inductive Learning (XHAIL) [14–16]. To date,
ILP has been applied to relatively small pathways, such as the AAA pathway
used in the early *Robot Scientist* work [7], while numerical methods have been
applied to much larger models, like the ABER model used in the subsequent
work [6]. While a previous XHAIL prototype was able to successfully revise the
pathway-specific AAA model in a variety of ways [14,16], scalability limitations
have prevented its application to the whole-organism ABER model until now.

This paper introduces a new implementation of XHAIL which scales from
AAA to ABER. Our main contributions are to overhaul the implementation
(from a single-threaded `Prolog` program to a multi-threaded *Java* application),
to upgrade the inference engine (from the outdated `Smodels` system to the state-
of-the-art `Clasp` solver), to support an extended language bias (which allows
the weighting of both examples and mode declarations), to revise the logical
encoding of metabolism (to better handle cellular compartments), to explore
the scalability of the system (with respect to the number of examples, reactions

© Springer International Publishing Switzerland 2015
J. Davis and J. Ramon (Eds.): ILP 2014, LNAI 9046, pp. 33–48, 2015.
DOI: 10.1007/978-3-319-23708-4_3

and constraints), to make the new system open-source[1] (under GPLv3), and to use our system to find an error in ABER (and in its influential precursor).

2   Metabolic Network Modelling

Metabolic networks are collections of interconnected biochemical reactions that mediate the synthesis and breakdown of essential compounds within the cell [8]. These reactions are catalysed by specific enzymes whose amino acid sequences are specified in regions of the host genome called Open Reading Frames (ORFs). Particular pathways within a network are controlled by regulating the expression of those genes on which the associated ORFs are located.

Much effort has been invested in the modelling of metabolic networks, leading to numerous qualitative reaction databases (such as the Kyoto Encyclopaedia of Genes and Genomes – KEGG [12]) and quantitative in-silico models (such as iFF708 [1]). Such resources fall into two classes: highly detailed *pathway-specific* models of small reaction sequences called pathways (often curated by hand); and less detailed *whole-organism* models that synthesise information about all known reactions (typically by computerised data mining of the literature).

These complementary trends are exemplified by a project known as the *Robot Scientist* [6,7] which integrates laboratory robotics and AI to automate high throughput growth experiments with the yeast *S. cerevisiæ*. The first phase of this project [7] used a pathway-specific model called AAA (of Aromatic Amino Acid bio-synthesis) while the second phase [6] used a whole-organism model called ABER (made at Aberystwyth University), as described below.

2.1   The AAA network

The AAA model is shown in Fig. 1. Nodes in the graph denote metabolites involved in the conversion of the start compound *Glycerate-2-phosphate* to the amino acids *Tyrosine*, *Phenylalanine*, and *Tryptophan*. Arrows denote chemical reactions from substrates to products. For convenience, nodes are annotated with their KEGG identifiers (in red); and arrows are labelled with 4-part EC numbers (in blue) and a set of ORF identifiers (in green).

ORFs appearing above each other denote iso-enzymes which all independently catalyse the same reaction (e.g., YHR137W and YGL202W), while ORFs appearing side by side denote enzyme-complexes built from several gene products that must be present simultaneously (e.g., YER090W and YKL211C). The dashed brown line shows the inhibition of an enzyme (YBR249C) by a metabolite (C00082). All reactions take place in the cell cytosol using nutrients imported from the growth medium; and they proceed at a standard rate, except for the import of two underlined compounds (C01179 and C00166), which take longer.

Although this model is based on data in KEGG, its authors manually included extra information from the literature relevant to growth experiments

---

[1] https://github.com/cathexis-bris-ac-uk/XHAIL.

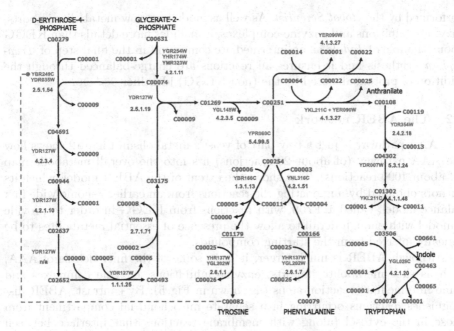

Fig. 1. Pathway-specific metabolic network represented by the AAA model.

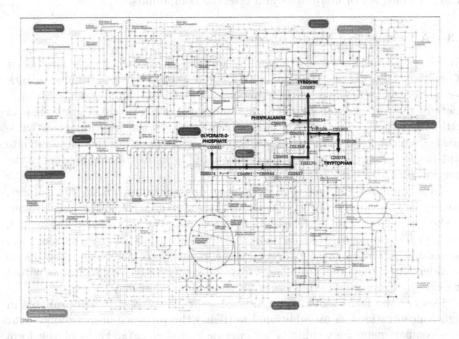

Fig. 2. Whole-organism metabolic map of yeast *S. cerevisiæ* with AAA pathway in black (adapted from an image at http://www.genome.jp © Kanehisa Laboratories) (Color figure online).

performed by the *Robot Scientist*. As well as modelling slow metabolite imports, enzyme-inhibitions and enzyme-complexes, it includes more details than KEGG about the role of *Indole* as an intermediate compound in the final step of *Tryptophan* synthesis and it ensures all reactions are charge-balanced through the addition of protons denoted by the (non-KEGG) identifier C00000.

2.2   The ABER network

The AAA pathway is just a tiny part of yeast's metabolism. Figure 2 shows how the AAA pathway (of about 20 reactions) fits into the overall metabolic map (of about 1000 reactions). This shows the extent of the ABER model, which its authors obtained by augmenting the reactions from an earlier genome-wide flux balance model, called iFF708, with reactions from KEGG in order to provide a model with enough detail to allow the presence of essential metabolites to be logically inferred from the starting compounds.

Although ABER is much larger, it lacks some of the finer details of AAA, such as slow metabolite imports, enzyme inhibitions, enzyme complexes and some intermediate reaction steps (as shown in Fig. 5). By contrast, ABER distinguishes reactions occurring in a separate mitochondrial compartment from those in the cytosol (along with membrane reactions that interface between them and the external growth medium). The ABER model also comes with a clearly defined set of ubiquitous and essential compounds.

2.3   A Declarative Model of Metabolism

To reason about the above networks we translated each graph into a set of ground facts in a common ASP representation with a unique identifier for every *metabolite*, ORF, *reaction, compartment* and enzyme *complex*. Facts of the form component(orf,cplx) and catalyst(rctn,cplx) state that an ORF orf is part of a complex cplx which catalyses the reaction rctn; while substrate(mtbl,cmpt,rctn) or product(mtbl,cmpt,rctn) state that a metabolite mtbl (located in a particular compartment cmpt) is a substrate or product of the reaction rctn.

To account for the fact that some reactions are more certain than others, our representation allows reactions to be flagged as *assertable* (i.e. initially out of the model, but can be included) or *retractable* (i.e. initially in the model, but can be excluded). This is done by declaring a set of facts of the form assertable_reaction(rctn) or retractable_reaction(rctn) which allow us to consider likely reactions from known reference pathways for inclusion within or exclusion from a revised network for a specific organism.

Facts of the form inhibitor(cplx,mtbl,cmpt) state that (when available in high concentrations as an additional nutrient) the presence of metabolite mtbl in a compartment cmpt inhibits an enzyme complex cplx. Facts of the form start_compound(mtbl,cmpt) and ubiquitous_compound(mtbl,cmpt) indicate that mtbl is always present in cmpt; while essential_compound(mtbl,cmpt) states that the presence of mtbl in cmpt is essential for cell growth.

We also encode the experimental results which indicate how yeast strains from which certain ORFs have been *knocked out* proliferate over a period of two days in growth media to which particular nutrients have been *added*. Each such experiment is given a unique identifier expt and the conditions are recorded by (zero or more) facts of the form additional_nutrient(expt,mtbl,cmpt) and knockout(expt,orf). The outcome on some day day is given by a literal of the form predicted_growth(expt,day) or not predicted_growth(expt,day).

At the heart of our approach is a set of rules for deciding which metabolites are in which compartments on which days in which experiments. These were taken from [14,16] subject to a few changes: we used predicate predicted_growth in our examples instead of using another predicate observed_growth forced to have the same extent by integrity constraints with classical negation; we made a distinction between ubiquitous_compound and start_compound; we added an extra argument into additional_nutrient, essential_compound and inhibitor to localise them all to a given compartment; we added an extra argument into inhibited to localise its effect to a given day[2]; and we modelled slow imports by inhibiting the relevant import reactions on the first day of each experiment. As these changes are small, we refer the reader to [14,16] for more details.

In brief, predicting the outcome of a given experiment amounts to checking if all essential metabolites are present in their required compartments. Apart from start compounds, ubiquitous compounds and additional nutrients, the only way a metabolite can be in a compartment is if it gets there as the product of an active reaction – whose substrates are all in their respective compartments, which has not been excluded (if it is retractable) or has been included (if it is assertable), and which is catalysed by a complex which is not inhibited on the day in question, and which has not been deleted (by having any of its component ORFs knocked out). This is represented by a set of ground rules like the following (which is one of two rules produced for reaction 2.5.1.19, assuming it has been given the identifier 31 and a retractable status):

```
1    in compartment(Expt,"C01269",cytosol,Day) :-
2        in compartment(Expt,"C00074",cytosol,Day),
3        in compartment(Expt,"C03175",cytosol,Day),
4        catalyst(31,Cplx),
5        not inhibited(Expt,Cplx,Day),
6        not deleted(Expt,Cplx),
7        not exclude(31).
```

3   Metabolic Network Revision

The fact that all the models we have mentioned so far (ABER, AAA, KEGG and iFF708) differ in the details of even just the tiny pathway shown above, and the fact that none of them fully agrees with experimental data, suggests that

---

[2] In fact this change was already present in the model used in [16] but the description in the paper incorrectly reproduced an earlier version of the rule from [14].

they are all approximations which are incomplete and/or incorrect. This raises the question of how such networks can be usefully revised (as was done manually in the creation of ABER and AAA). Prior work has shown that ILP can help to automate this process. In particular, the systems `Progol5` and XHAIL have been used to complete and revise the AAA model, as described below.

3.1   ILP

This paper assumes a familiarity with Inductive Logic Programming (ILP) [11]. An atom a is a predicate p applied to a collection of terms $(t_1, \ldots, t_n)$. A literal l is an atom a (positive literal) or the negation of an atom $not\ a$ (negative literal). A clause C is an expression of the form $a :- l_1, \ldots, l_m$ where a is an atom (head atom) and the l_i are literals (body literals). A clause with no body literals is called a fact and a clause with no head atom is called a constraint. A logic program P is a set of clauses. In this paper, the meaning of a logic program is given by the Answer Set or Stable Model semantics [3]. A program P entails a set of ground literals L, denoted $P \models L$, if there is an answer set of P that satisfies all of the literals l in L. A logic program H entails a set of ground literals E with respect to a logic program B if $B \cup H \models E$.

The task of ILP is to find a hypothesis H that entails some examples E with respect to a background theory B and also satisfies some form of language and search *bias*. The search bias typically used is that of minimising *compression* [9] of the hypothesis with respect to the examples (where compression is the number of literals in the hypothesis minus the number of examples covered). The language bias typically used is that of specifying a set of *mode declarations* [9] that allow the user to impose various constraints on which literals may appear in the heads and bodies of hypothesis clauses (as illustrated later).

3.2   Progol5

`Progol5` [9,10] is a popular ILP system which was used in the first phase of the *Robot Scientist* work to extend the AAA model with missing ORF-enzyme mappings [7]. It uses a greedy covering approach that explains one example at a time by constructing and generalising a ground clause called a *Bottom Set* (BS) that efficiently bounds the search space. First the head of the BS is computed by a contrapositive method, then the body of the BS is computed by a saturation method, and finally the BS is generalised by a subsumption search.

While `Progol5` has been used to monotonically extend AAA with missing information, its limited ability to reason abductively or inductively through negation means it is not well-suited for tackling revision problems – which are inherently nonmonotonic. Its greedy covering approach assumes the addition of a future hypothesis cannot undermine the truth of an earlier example and, even in this limited setting, it is (semantically and procedurally) incomplete [13].

3.3   XHAIL

eXtended Hybrid Abductive Inductive Learning(XHAIL) [15] overcomes the incompleteness of `Progol5` (at the expense of solving a harder problem). For soundness it does not process examples one-by-one, but solves them all at once. For completeness it does not generalise a single BS, but a *set* of such clauses, K, known as a *Kernel Set*(KS). For efficiency it uses iterative deepening to consider progressively larger Kernel Set.

Existing Prototype While its semantics is defined using the *Generalised Stable Model* (GSM) notion from *Abductive Logic Programming* (ALP) [5], the existing XHAIL prototype uses *Answer Set Programming* (ASP) [3] to perform the three key phases of the algorithm. Firstly, the head atoms of K are computed *abductively* to give a set of ground atoms Δ that entail E with respect to B. Then, the body atoms of K are computed *deductively* to give a set of ground atoms Γ entailed by $B \cup \Delta$. Finally, K is generalised *inductively* to give a compressive hypothesis H subsuming K.

As described in [16] the existing prototype, which used `Smodels` as the ASP system [17], was able to successfully revise AAA to make it consistent with the *Robot Scientist* data from [7] including a number of results known to contradict the final AAA model (as noted in *Section 3* of the *supplementary material* for [7]). Unfortunately, scalability limitations meant the existing prototype could not be used on the ABER. Hence, in this paper one of our key aims was to develop a new implementation of XHAIL that can be applied to ABER.

New Architecture To make our new system portable, extensible and efficient, we wrote it from scratch in *Java 8* using a modular architecture that exploits multi-threading and has been made open-source under *GNU General Public License v3.0* (GPLv3). We also upgraded the ASP engine from the (now unsupported) `Smodels` to the (current ASP competition winner) `Clasp` [2] and, to make the system more usable and noise-tolerant, we also introduced an extended language bias that allows mode declarations and examples to be weighted.

An overview of the system is given in Fig. 3. The `Application` class acts as the user interface. When invoked, it generates an instance of a `Problem` and activates a bespoke `Parser` (which can recognise both `Clasp` and XHAIL-specific statements) to load the desired knowledge base.

The `solve` method in `Problem` triggers the procedure by invoking a `Dialler` on an instance of the `Problem`. The `Dialler` calls `Clasp` and collects the results in a temporary file (interpreted by a specialised parser `Acquirer`) to return the set of solutions as a `Result`. The `Problem` and `Clasp` are loosely-coupled so they can be invoked as separate threads to achieve more parallelism. Now, at the end of the (*abductive*) stage, the `Result` contains the head atoms Δ of the KS K.

An instance of `Grounding` is then used to *deductively* derive the body atoms Γ of K. Its `solve` method is called to activate the `Dialler` on the current `Grounding` to perform *induction*. The `Result` read at this point from the temporary file is used to generate as many `Hypotheses` as needed, each representing

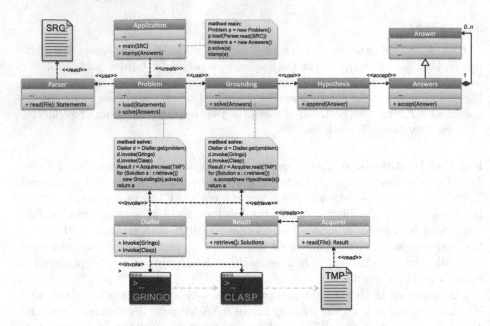

Fig. 3. Architectural overview of our new XHAIL implementation.

an *H*. Each Hypothesis found in this fashion is converted into an Answer and appended to the set of Answers. If any weights have been applied to examples or mode declarations, each Answer will have a score, and any non-optimal ones will be discarded. Finally, the Application presents the Answers to the user.

New Language Bias As shown in Fig. 4 we have extended the new XHAIL's language bias to allow integer weights and priorities (familiar concepts in ASP) to be associated with individual examples and mode declarations. Adding a weight to an example has the effect of making that example defeasible (so the example can remain uncovered at the expense of paying a penalty set by the weight). Adding a weight to a mode declaration has the effect of incurring a penalty set by the weight every time the mode declaration is used.

The overall effect of these weights is to give preference to certain hypotheses over others; and the task of the new XHAIL is to find hypotheses that incur the smallest penalty. We believe this offers the user a finer level of control than the default compression metric. Of course, the standard bounds on the maximum and minimum number of times a mode declaration can be used are still available.

4   Case Study

To validate the approach, we applied our new XHAIL system to our logical encoding of the ABER model. This was done in two stages. The first stage was to revise the ABER model to incorporate the extra information in the more

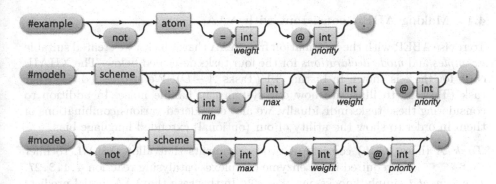

Fig. 4. Syntax diagrams for the new XHAIL's extended language bias.

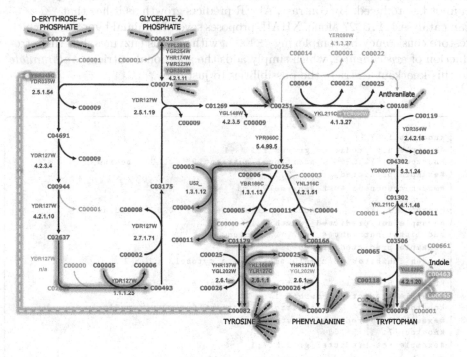

Fig. 5. Fragment of ABER corresponding to AAA pathway with missing information in grey and other changes highlighted in orange and purple (Color figure online).

detailed AAA model. The second stage was to learn some additional hypotheses (originally found in [16]) to make the model consistent with real *Robot Scientist* data. In both cases, we studied how the approach scaled to progressively bigger fragments of the model with increasingly large sets of observations. We began with the fragment shown in Fig. 5, where the purple highlights show the four pieces of information we wanted to learn in the first validation phase (to make this sub-network of ABER functionally equivalent to AAA).

4.1   Making ABER consistent with AAA

To revise ABER with the information from AAA that it lacks, we created suitable *examples* and *mode declarations* for the four tasks described below. The XHAIL code for these is shown in Listing 3.1 (Tasks $A - D$). We also added a further task (Task E) to illustrate how our system can handle noise. In addition to considering these tasks individually, we also considered various combinations of them in order to show the utility of our (optional) extended language bias.

Task A (lines $1 - 6$) consists of two experiments that allow XHAIL to infer that YER090W is required in all enzyme complexes catalysing reaction 4.1.3.27. *Experiment 1* simply knocks out YER090W. In this case the AAA model predicts no growth since all enzymes producing *Anthranilate* are blocked and *Tryptophan* cannot be produced. By contrast, ABER predicts growth as it has that YKL211C can catalyse 4.1.3.27 alone. XHAIL proposes several minimal hypotheses which restore consistency by complexing YER090W with various enzymes, but the introduction of *experiment 2*, which simply adds the additional nutrient *Anthranilate* to this knockout restricts the possibilities to just one, YKL211C.

```
 1  % Task A: YER090W as enzyme complex in 4.1.3.27
 2  knockout(1,"YER090W").
 3  #example not predicted_growth(1,1)  =4 .
 4  knockout(2,"YER090W").additional_nutrient(2,"C00108",medium).
 5  #example predicted_growth(2,1)  =4 .
 6  #modeh component(#orf,#enzymeID)  :1  =3 .
 7  % Task B: C00082 inhibits YBR249C in 2.5.1.54
 8  knockout(3,"YDR035W").additional_nutrient(3,"C00082",medium).
 9  #example not predicted_growth(3,1)  =4 .
10  additional_nutrient(4,"C00082",medium).
11  #example predicted_growth(4,1)  =4 .
12  #modeh inhibitor(#enzymeID,#metabolite,cytosol)  :1  =2 .
13  % Task C: C00463 contamination in 4.2.1.20
14  knockout(5,"YKL211C").additional_nutrient(5,"C00463",medium).
15  #example predicted_growth(5,1)  =4 .
16  knockout(6,"YGL026C").additional_nutrient(6,"C00463",medium).
17  #example not predicted_growth(6,1)  =4 .
18  knockout(7,"YKL211C").
19  #example not predicted_growth(7,1)  =4 .
20  #modeh include(#assertable_reaction)  :1  =2 .
21  #modeh catalyst(#assertable_reaction,#enzymeID)  :1  =1 .
22  % Task D: slow import of C00166 and C01179
23  knockout(8,"YBR166C").additional_nutrient(8,"C01179",medium).
24  #example not predicted_growth(8,1)  =4 .
25  knockout(9,"YNL316C").additional_nutrient(9,"C00166",medium).
26  #example not predicted_growth(9,1)  =4 .
27  #example predicted_growth(10,1)  =4 .
28  #modeh inhibited(+experiment,#enzymeID,#day)  :2  =1 .
29  % Task E: defeasible example
30  #example not predicted_growth(11,1)  =4 .
```

Listing 3.1. *Tasks A-E* with optional weights and constraints highlighted.

Table 1. Execution time (and number of hypotheses) vs. number of tasks and reactions as optional language bias is excluded (top half, shaded blue) or included (bottom half, shaded green).

Size	Task A		Task B		Task C		Task D		Task E	
52	0.3	(1)	0.4	(1)	0.3	(1)	7.2	(1)	0.166	(0)
87	0.5	(1)	0.7	(1)	0.7	(1)	10.5	(1)	0.214	(0)
127	0.8	(1)	1.9	(1)	1.5	(1)	18.7	(1)	0.294	(0)
207	1.8	(1)	3.9	(1)	2.3	(1)	39.7	(1)	0.449	(0)
367	4.8	(1)	6.0	(1)	3.9	(1)	76.3	(1)	0.814	(0)
687	14.1	(1)	15.6	(1)	9.5	(1)	186.0	(1)	1.802	(0)
1106	38.3	(1)	59.7	(1)	17.6	(1)	339.1	(0)	3.898	(0)
52	0.3	(1)	0.3	(1)	0.3	(1)	7.5	(1)	0.169	(1)
87	0.5	(1)	0.5	(1)	0.4	(1)	11.0	(1)	0.22	(1)
127	0.8	(1)	1.0	(1)	0.6	(1)	19.4	(1)	0.295	(1)
207	1.8	(1)	2.0	(1)	1.1	(1)	39.9	(1)	0.44	(1)
367	5.2	(1)	5.7	(1)	2.5	(1)	72.5	(1)	0.793	(1)
687	14.2	(1)	13.7	(1)	5.8	(1)	180.4	(1)	1.726	(1)
1106	38.4	(1)	29.4	(1)	14.7	(1)	342.8	(181)	3.827	(1)

Task B (lines 7 − 12) consists of two experiments that allow XHAIL to infer that *Tyrosine*, when used as an additional nutrient, inhibits the enzyme YBR249C catalysing reaction 2.5.1.54. *Experiment 3* simply knocks out the other enzyme YDR035W catalysing this reaction and adds the nutrient *Tyrosine*. While AAA predicts no growth since the only available enzyme for 2.5.1.54 is inhibited, ABER predicts growth as it has no information about inhibition. XHAIL again proposes several hypotheses which restore consistency by inhibiting various enzymes, but the addition of *experiment 4*, which simply adds *Tyrosine* to the wild type restricts the possibilities to just the intended one, YBR249C.

Tasks A and B (lines 1 − 12) can both be given to XHAIL at the same time. In this case it successfully learns the union of the two individual hypotheses, but it also learns several other hypotheses resulting from the interaction of the two tasks. The additional hypotheses all introduce two enzyme complexes, as opposed to one enzyme complex and one inhibitor. Instead of adding further examples to eliminate the unwanted hypotheses, we can exploit the ability of our new system to specify weights on mode declarations. By giving a lower cost (=2) to the mode declaration for inhibitors (line 12) than the cost (=3) for the mode declaration for complex components (line 6), we ensure only the intended hypothesis has minimal cost and is returned by XHAIL. Alternatively we can set the number of times each mode declaration is used to exactly once (:1). As shown below, such biases reduce the search space and the execution time.

Tasks C and D (lines 13–28) are similarly designed to rediscover a missing reaction for C00463 along with the slow import of C00166 and C01179.

Task E was also added as an obviously incorrect example (as it simply asserts that the wild type does not grow under normal conditions – which is false). Ordinarily, this would result in logical inconsistency, but by exploiting the ability of XHAIL to give defeasible weights to *examples*, we can still learn all the intended changes when running *tasks E* by paying a simple cost (=4) associated with this last example. Although the inconsistent *example* is discarded, XHAIL finds it is cheaper to construct hypotheses that explain all the other examples (as opposed to paying the penalty for discarding them).

Table 1 shows the time taken to run these tasks on various fragments of the ABER model. The x-axis shows the 5 *tasks* $A - E$. The y-axis shows the increasing sizes of model we used, starting with the initial network in Fig. 5 (which has 52 reactions including imports) all the way up to the whole model (with $1,106$ reactions). Experiments were run on a MacBook Pro featuring a 2.3 GHz Intel Core i7 with 4 cores and 16 GB DDR3 RAM.

The top half of the table refers to experiments using simple *mode declarations* without number constraints or weights, while the bottom half refers to experiments using our enhanced *mode declarations* with number constraints and weights. Each cell contains the execution time in seconds together with the number of answers returned (in parentheses). These tests show that the experiments become more computationally intensive as the number of reactions and tasks increases; and they also show that the language bias effectively helps to narrow down the number of answers and to reduce the execution times.

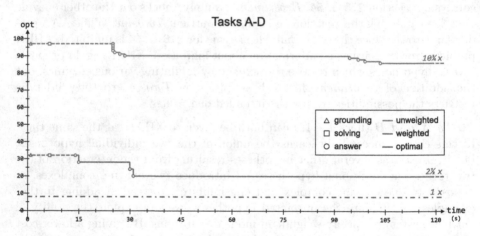

Fig. 6. Elapsed times vs. scores of answers found during resolution with language bias excluded (in red) or included (in green) with respect to the optimal (in blue) (Color figure online).

Figure 6 describes what happens during the first 2 minutes of execution when XHAIL tackles a problem, specifically all the *tasks* $A - D$ simultaneously run on the ABER fragment in Fig. 5. XHAIL still finds the expected hypotheses, but also several others resulting from the interaction between the *tasks*, making it a much harder problem to solve. In order to eliminate the unwanted hypotheses and show the utility of our extended language bias, we have introduced further constraints and weights on the mode declarations of the problem (as highlighted in yellow in Listing 3.1). This graph compares the optimality score and the elapsed time of each answer found when the optional language bias is excluded (in red) or included (in green) with respect to the score expected for the optimal solution (in blue). The triangles indicate when the parsing and generation of the abductive phase problem terminates and `Gringo`'s grounding starts. Similarly, the squares show when the grounding is over and `Clasp` takes over. Finally the circles show plot the improvement in the score of the current best solution (y-axis) against time (x-axis). The lines preceding the first answers are dashed because at that point is not possible to know in advance the first optimality value but it is convenient to show where triangles and squares sit. Notice how the extended language bias results in higher quality answers being found in less time.

4.2   Making ABER consistent with *Robot Scientist* data

We used the revised ABER model to run 40 additional experiments for which growth data had been previously obtained by a *Robot Scientist*. We set the mode declarations as follows (where the highlighted expressions :min-max denote optional bias which constrains how many times each mode declaration is used):

```
1  #modeh inhibited(+experiment,#modifiable_enzyme,#day)  :0-2 .
2  #modeh additional_nutrient(+experiment,#metabolite,medium)  :0-1 .
3  #modeb additional_nutrient(+experiment,#metabolite,medium)  :0-1 .
```

This allows XHAIL to learn the hypothesis from [16] consisting of the clause "`additional_nutrient`(E,`"C00078"`,`medium`) :- `additional_nutrient`(E, `"C00463"`, `medium`)" – which suggests the contamination of *Indole* with *Tryptophan* (as the latter becomes an additional nutrient whenever the former is) – and a fact "`inhibited`(E,10,1)" – which suggests the slow import of *Anthranilate* (as the hypothetical enzyme catalysing its import is inhibited on the first day of each experiment). The time taken to run more and more experiments on larger

Table 2. Execution time (and number of hypotheses) vs. number of experiments and reactions as optional language bias is excluded (top half, shaded blue) or included (bottom half, shaded green).

Size	Ex 1-4		Ex 1-8		Ex 1-16		Ex 1-32		Ex 1-40	
52	0.2	(1)	0.8	(1)	3.6	(1)	19.1	(1)	65.8	(10)
87	0.5	(1)	3.5	(1)	18.3	(1)	110.2	(1)	≥600	(≥1)
127	0.8	(1)	7.1	(1)	64.5	(1)	510.3	(1)	≥600	(≥1)
207	1.8	(1)	12.5	(1)	80.0	(1)	≥600	(≥1)	≥600	(≥1)
367	3.5	(1)	26.3	(1)	265.0	(1)	≥600	(≥1)	≥600	(≥1)
687	8.1	(1)	101.9	(0)	≥600	(≥1)	≥600	(≥1)	≥600	(≥1)
1106	14.5	(1)	262.0	(0)	≥600	(≥1)	≥600	(≥1)	≥600	(≥1)
52	0.2	(1)	0.6	(1)	1.5	(1)	5.8	(1)	11.4	(10)
87	0.5	(1)	1.6	(1)	5.4	(1)	14.5	(1)	40.0	(10)
127	0.8	(1)	3.5	(1)	12.9	(1)	43.4	(1)	202.4	(10)
207	1.8	(1)	8.7	(1)	42.0	(1)	234.4	(1)	440.1	(10)
367	3.9	(1)	19.5	(1)	≥600	(≥1)	≥600	(≥1)	≥600	(≥1)
687	8.3	(1)	110.1	(24)	531.8	(4)	≥600	(≥1)	≥600	(≥1)
1106	15.6	(1)	300.5	(24)	≥600	(≥1)	≥600	(≥1)	≥600	(≥1)

and larger networks is shown in Table 2, which is formatted like Table 1 with the shading of each cell increasing with execution time.

It is interesting to note that the number of solutions drops from 1 to 0 when experiments 1 − 8 are run on the fragment of ABER with 687 reactions (without the optional bias). We suspected one of the additional reactions was interfering with the initial AAA fragment so we added an extra mode declaration to allow XHAIL to exclude zero or more reactions. This led to a hypothesis which excluded one reaction with EC 1.3.1.12 and ORF U52_ which, although it is shown in Fig. 5, was in fact accidentally left out of the initial fragment because we did not notice it when carrying out our initial analysis by hand.

Since we could not find any hypothesis in which that reaction was not excluded, we examined the literature support for this reaction and found that, while it does exist in the KEGG reference pathway, it does not exist in any of the pathways specific to yeast. We then determined that it was imported into ABER from iFF708. In this way, we suggest our method has led to the discovery of a mistake in both ABER and iFF708.

5   Comparison with Progol5

It is interesting to see why the Progol5 system used in the early *Robot Scientist* work, cannot be used on tasks like those solved in Sect. 4. For this purpose, it suffices to consider the following Progol5 input file which simplifies the task of including and excluding reactions to its absolute bare principles:

```
 1  % Background Knowledge
 2  product(m,ass).
 3  product(n,ret).
 4  present(MN) :- product(MN,ass), include(ass).
 5  present(MN) :- product(MN,ret), not exclude(ret).
 6  % Examples
 7  :- observable(present/1)?
 8  present(m).
 9  :- present(n).
10  % Language Bias
11  :- modeh(1,include(ass))?
12  :- modeh(1,exclude(ret))?
```

The background theory declares a metabolite m as the product of an assertable reaction ass and a metabolite n as the product of a retractable reaction ret. It then states the product of an assertable reaction will be present if the reaction is included; while the product of a retractable will be present if the reaction is *not* excluded. The bias allows Progol5 to include an assertable reaction and/or to exclude a retractable reaction.

While XHAIL easily learns the hypothesis include(ass) and exclude(ret) from these inputs, Progol5 simply states that this input is inconsistent and it fails to learn anything. This is because the example atom present(m) depends negatively upon the intended hypothesis exclude(ret), which is therefore not amenable to Progol5's limited form reasoning with negation.

It might seem Progol5 can be made to find the answer by transforming this nonmonotonic problem into a monotonic one by reifying the literals include(ass) and not exclude(ret) to the atoms include(ass,true) and exclude(ret,false) – while adding the following constraints to avoid meaningless solutions where a reaction is both included and not included at the same time (which is the very approach taken in another Progol5 application called Metalog [18]):

```
1  :- include(R,true),include(R,false).
2  :- exclude(R,true),exclude(R,false).
```

Although Progol5 can now learn one of the hypotheses include(ass,true), this has the semantically problematic result of requiring the exclusion of the retractable reaction to be neither true nor false. And this is unavoidable as adding the following complementary constraints to force the issue merely re-establishes the logical inconsistency that the reification was introduced to avoid:

```
1  :- not include(R,true), not include(R,false).
2  :- not exclude(R,true), not exclude(R,false).
```

It can be shown that any semantically meaningful solution to this problem will require explicitly formalising the rules to determine when a compound is *not* in a compartment – which is much harder than formalising when a compound is in a compartment and is not practical in more realistic versions of this task.

6 Conclusions

This paper introduced a new implementation of XHAIL that is able to usefully revise a whole-organism model of yeast metabolism. It summarised some of the key implementation improvements and investigated their scalability with respect to the number of examples and the size of the network. It also introduced some new language bias extensions and analysed their impact on execution times and the number of hypotheses returned. The new system is much faster than the old prototype and can be applied to much larger networks.

To the best of our knowledge, this is the first application of ILP to a whole-organism model of metabolism; and it has led to the discovery of a reaction that we believe is incorrectly included in both ABER and its influential precursor iFF708 (which is the forerunner of most whole-organism flux-balance models of yeast developed to this day). These preliminary results suggest that XHAIL is a useful method for biological network revision (and probably for other non-monotonic ILP tasks as well). Both the system and all of the input files used in this work have been made fully open-source under GPLv3.

In future work we plan to investigate in more detail how our execution times are distributed across the different phases (adductive, deductive and inductive) of XHAIL and to exploit a parallel implementation of Clasp to help further speed up execution of our system. We also need to test the system in harder case studies where the required language bias is not set in advance and see if we can apply our method to more recent models of yeast metabolism, such as Yeast5 [4], which are not designed to be logically executable.

Acknowledgments. This work is supported by EPSRC grant *EP/K035959/1*.

References

1. Förster, J., Famili, I., Fu, P., Palsson, B., Nielsen, J.: Genome-scale reconstruction of the saccharomyces cerevisiae metabolic network. Gen. Res. **13**(2), 244–53 (2003)
2. Gebser, M., Kaufmann, B., Neumann, A., Schaub, T.: clasp: A conflict-driven answer set solver. In: Logic Programming and Nonmonotonic Reasoning, pp. 260–265. Springer (2007)
3. Gelfond, M., Lifschitz, V.: Classical negation in logic programs and disjunctive databases. New Generation Comp. **9**(3/4), 365–386 (1991)
4. Heavner, B.D., et al.: Yeast 5 – an expanded reconstruction of the saccharomyces cerevisiae metabolic network. BMC Syst. Biol. **6**, 55 (2012)
5. Kakas, A., Kowalski, R., Toni, F.: Abductive logic programming. J. Logic Comput. **2**(6), 719–770 (1992)
6. King, R.D., Rowland, J., Oliver, S.G., Young, M., Aubrey, W., Byrne, E., Liakata, M., Markham, M., Pir, P., Soldatova, L.N.: The automation of science. Science **324**(5923), 85–89 (2009)
7. King, R.D., Whelan, K.E., Jones, F.M., Reiser, P.G., Bryant, C.H., Muggleton, S.H., Kell, D.B., Oliver, S.G.: Functional genomic hypothesis generation and experimentation by a robot scientist. Nature **427**(6971), 247–252 (2004)
8. Lehninger, A.: Biochemistry: The Molecular Basis of Cell Structure and Function, 2nd edn. Worth Publishers, New York (1979)
9. Muggleton, S.: Inverse entailment and Progol. New Gen. Comp. **13**, 245–286 (1995)
10. Muggleton, S.H., Bryant, C.H.: Theory completion using inverse entailment. In: Cussens, J., Frisch, A.M. (eds.) ILP 2000. LNCS (LNAI), vol. 1866, pp. 130–146. Springer, Heidelberg (2000)
11. Muggleton, S., De Raedt, L.: Inductive logic programming: theory and methods. J. Logic Program. **19**(20), 629–679 (1994)
12. Ogata, H., Goto, S., Sato, K., Fujibuchi, W., Bono, H., Kanehisa, M.: Kegg: Kyoto encyclopedia of genes and genomes. Nucleic Acids Res. **27**(1), 29–34 (1999)
13. Ray, O.: Hybrid Abductive-Inductive Learning. Ph.D. thesis, Department of Computing, Imperial College London, UK (2005)
14. Ray, O., Whelan, K., King, R.: A nonmonotonic logical approach for modelling and revising metabolic networks. In: Proceedings of 3rd International Conference on Complex, Intelligent and Software Intensive Systems, pp. 825–829. IEEE (2009)
15. Ray, O.: Nonmonotonic abductive inductive learning. J. Appl. Logic **7**(3), 329–340 (2009)
16. Ray, O., Whelan, K., King, R.: Automatic revision of metabolic networks through logical analysis of experimental data. In: De Raedt, L. (ed.) ILP 2009. LNCS, vol. 5989, pp. 194–201. Springer, Heidelberg (2010)
17. Simons, P., Niemelä, I., Soininen, T.: Extending and implementing the stable model semantics. Artificial Intel. **138**(1–2), 181–234 (2002)
18. Tamaddoni-Nezhad, A., Kakas, A.C., Muggleton, S.H., Pazos, F.: Modelling inhibition in metabolic pathways through abduction and induction. In: Camacho, R., King, R., Srinivasan, A. (eds.) ILP 2004. LNCS (LNAI), vol. 3194, pp. 305–322. Springer, Heidelberg (2004)

Construction of Complex Aggregates
with Random Restart Hill-Climbing

Clément Charnay[✉], Nicolas Lachiche, and Agnès Braud

ICube, Université de Strasbourg, CNRS, 300 Bd Sébastien Brant,
CS 10413, 67412 Illkirch Cedex, France
{charnay,nicolas.lachiche,agnes.braud}@unistra.fr

Abstract. This paper presents the integration of complex aggregates in
the construction of logical decision trees to address relational data mining tasks. Indeed, relational data mining implies aggregating properties of
objects from secondary tables and complex aggregates are an expressive
way to do so. However, the size of their search space is combinatorial and
it cannot be explored exhaustively. This leads us to introduce a new algorithm to build relevant complex aggregate features. This algorithm uses
random restart hill-climbing to build complex aggregation conditions.
The algorithm shows good results on both artificial data and real-world
data.

1 Introduction

Relational data mining deals with data represented by several tables. We focus
on the typical setting where one table, the primary table, contains the target
column, and has a one-to-many relationship with a secondary table. A possible
way of handling such relationship is to use complex aggregates, i.e. to aggregate the objects of the secondary table which verify a given condition, using an
aggregate function on the objects themselves (*count* function) or on a numerical
attribute of the objects (e.g. *max*, *mean* functions).

Previous work showed that the expressivity of complex aggregates can be
useful to solve problems such as urban blocks classification. [6,9] introduced
complex aggregates in propositionalisation. But their use increases the size of
the feature space too much, and they cannot be fully handled. This is the reason
why we focus on introducing them in the learning step. The main motivation for
our work is to elaborate new heuristics to handle complex aggregates in relational
models.

This article presents a logical decision tree learner which uses complex aggregates to deal with secondary tables, using a random restart hill-climbing heuristic
to build them. This heuristic allows, for a fixed couple aggregation function and
aggregated attribute, to find a complex aggregation condition to select the relevant objects to aggregate from the secondary table. We introduce this heuristic in
the context of logical decision trees, but it could be applied to other approaches.

The rest of this paper is organized as follows. Section 2 introduces useful notations. Section 3 is a review of the use of aggregates in relational learners. Section 4

© Springer International Publishing Switzerland 2015
J. Davis and J. Ramon (Eds.): ILP 2014, LNAI 9046, pp. 49–61, 2015.
DOI: 10.1007/978-3-319-23708-4_4

describes our heuristic to explore the complex aggregate space. Section 5 shows experimental results. Finally, Sect. 6 concludes and suggests future work.

2 Preliminaries

This section introduces notations about relational databases and complex aggregates. We then provide an illustrative example.

2.1 Notations

Relational Database. The notations for relational databases are inspired by those introduced in [4]. A relational database, noted D, is seen as a set of N tables $D = \{T_1, T_2, \ldots, T_N\}$.

A table T has :

- At least one primary key attribute, denoted by T.K.
- Descriptive attributes, their number is denoted by $arity(T)$, their set is denoted by $T.A = \{T.A_1, T.A_2, T.A_{arity(T)}\}$. The definition domain of a descriptive attribute is denoted by $domain(T.A_k)$, if the attribute is nominal, then it is a set, if it is numerical, then it is an interval of $\mathbb{R}$.
- Foreign keys, their number is denoted by $f(T)$, their set is denoted by $T.F = \{T.F_1, T.F_2, T.F_{f(T)}\}$. Each foreign key references the primary key of another table. The primary key referenced by the foreign key $T.F_k$ is denoted by $ref(T.F_k)$.

In supervised learning, the value of one descriptive attribute of one table is to be predicted. We will refer to this table as the main table, and to the attribute as the class attribute.

Complex Aggregates. In a setting with two tables linked through a one-to-many relationship, let us denote the main table by M and the secondary table by S, i.e. there exists an index k such that $S.F_k \in S.F$ and $ref(S.F_k) = M.K$. We define a complex aggregate feature of table M as a triple (*Selection, Feature, Function*) where:

- Selection selects the objects to aggregate. It is a conjunction of s conditions c, i.e. $Selection = \bigwedge_{1 \leq i \leq s} c_i$, where c_i is a condition on a descriptive attribute of the secondary table. Formally, let $Attr \in S.A$, then c_i is:
  - either $Attr \in vals$ with $vals$ a subset of $domain(Attr)$ if $Attr$ is a categorical feature,
  - or $Attr \in [val_1; val_2[$ with $val_1 \in domain(Attr)$, $val_2 \in domain(Attr)$ and $val_2 > val_1$ if $Attr$ is a numerical feature.
  In other words, for a given object of the main table, the objects of the secondary table that meet the conditions in *Selection* are selected for aggregation.

– *Feature* can be:
  • nothing,
  • a descriptive attribute of the secondary table, i.e. *Feature* ∈ *S.A.*
Thus, *Feature* is the attribute of the selected objects that will be aggregated.
It can be nothing since the selected objects can simply be counted, in which
case a feature to aggregate is not needed.
– *Function* is the aggregation function to apply to the bag of feature values for
the selected objects. In this work we will consider six aggregation functions:
count, min, max, sum, mean and *standard deviation*.

In the rest of the paper, we will denote a complex aggregate by *Function(Feature, Selection)*.

2.2 Example: Urban Blocks

We introduce an example with two tables: the table of urban blocks and the table
of buildings. Those two tables are linked through a composition relationship,
which is in this relational setting a one-to-many relationship: one urban block is
linked to several buildings. The database is represented in Fig. 1.

block_id	density	convexity	elongation	area	class
1	0.151	0.986	0.221	22925	h_indiv
2	0.192	0.832	0.155	15363	h_coll
3	0.204	0.718	0.450	17329	h_mixed
...	...	...	...	...	...

Main table - blocks

1

0..N

building_id	convexity	elongation	area	block_id
1_1	1.000	0.538	165	1
1_2	0.798	0.736	323	1
1_3	1.000	0.668	84	1
...	...	...	...	...
2_1	0.947	0.925	202	2
2_2	1.000	0.676	147	2
...	...	...	...	...

Secondary table - buildings

Fig. 1. Schema of the real-world urban block dataset

The table of urban blocks has:

- A primary key *block_id*.
- Five descriptive attributes: the class attribute *class*, and 4 numeric attributes *density*, *convexity*, *elongation* and *area*.
- No foreign key.

The table of buildings has:

- A primary key *building_id*.
- Three descriptive attributes: *convexity*, *elongation* and *area*.
- One foreign key: *block_id* which references the table of urban blocks.

Let us give three examples of complex aggregate features, taken from the many possible ones:

- *count(buildings(area $\geq$ 450))* computes the number of buildings in the block with area greater than or equal to 450.
- *mean(elongation, buildings(area $\geq$ 450))* computes the average elongation of buildings in the block with area greater than or equal to 450.
- *max(convexity, buildings(area $\geq$ 450 $\wedge$ elongation < 0.7))* computes the maximum of miller of buildings in the block with area greater than or equal to 450 and elongation lower than 0.7.

3   Related Work

In this section, we review relational learners that take advantage of aggregation to build relational features.

RELAGGS [8], is a propositionalisation algorithm that aggregates the attributes of secondary tables, by using simple aggregation. Formally, the *Selection* part of the aggregate is empty: the algorithm does not use conditions to aggregate a subset of secondary objects, but aggregates them all. We intend to go further, by allowing the aggregation of a subset of objects.

TILDE [3] is a first-order decision tree learner. It uses a top-down approach to choose, node by node, from root to leaves, the best refinement to split the training examples according to their class values, using gain ratio as a metric to guide the search. TILDE relies on a language bias: the user specifies the shape of the literals which can be added in the conjunction at a node. In this relational context, to deal with secondary tables, the initial version of TILDE introduces new variables from these secondary tables using an existential quantifier.

Then, TILDE was extended [1] to allow the use of complex aggregates, and a heuristic has been developed to explore the search space [11]. This heuristic is based on the idea of a refinement cube, where the dimensions of the cube are refinement space for the aggregation condition, aggregation function and threshold. This cube is explored in a general-to-specific way, using monotone paths along the different dimensions: when a complex aggregate (a point in the refinement cube) is too specific (*i.e.* it fails for every training example), the search does not restart from this point.

However, the implementation does not allow more than two conjuncts in the aggregate query "due to memory problems" [2, p. 32], which limits the aggregation condition to one simple condition on an attribute of the secondary table. Our goal is as follows: to build a relational decision tree learner, which can use as node splits conditions on both attributes of the primary table and complex aggregates built from secondary tables, with no limitation on the size of the aggregation condition.

Use of stochastic optimization algorithms based on random restart for feature construction has been proposed by [7]. The aim of the authors is to find a relevant set of binary features for model construction. Although our procedure is similar to the basic randomised local search, our goal by using random restart hill-climbing is to find one relevant split, which can be seen as a binary feature, of the data in terms of information gain. This approach was also only applied to the construction of classical ILP features, based on the existential quantifier. It has never been applied to the construction of complex aggregates, whose expressivity increases the complexity of the feature space. Our procedure is different because it takes advantage of the structure of the complex aggregates search space, by combining several random restart hill-climbing processes.

4 Hill-Climbing of Complex Aggregate Conditions

In this section, we introduce a new algorithm based on random restart hill-climbing to generate relevant complex aggregates for the problem at hand.

Our algorithm considers one process of random restart hill-climbing per couple (*Function, Feature*), we will refer to such a process as a branch.[1] We do so because a change of function or feature to aggregate in the middle of a hill-climbing process is very drastic and would lead to a comparison between aggregates which should not be compared. As a result, the hill-climbing is only performed on the *Selection* part of the aggregate.

As introduced previously, the *Selection* part of the aggregate is a conjunction of conditions on attributes of the secondary table. The hill-climbing algorithm initializes *Selection* by selecting secondary attributes, each one having a probability $\frac{1}{size(S.A)}$ of being selected. The right hand side of the condition is then completed by picking a possible value of the categorical attribute with probability 0.5, or by choosing the bounds of the interval at random for a numerical attribute. In this last case, the interval can be unbounded on one side with probability 0.5 (0.25 for each configuration), and bounded on both sides with probability 0.5.

Once the condition is initialized, the hill-climbing algorithm will, at each step, try to apply refinement operations of the conjunction. These refinement operations are:

---

[1] This concept of branch has nothing to do with the concept of branch in a decision tree. We refer here to a part of the complex aggregate search space.

- Add a newly initialized condition on an attribute not present in the conjunction.
- Remove a condition on an attribute from the conjunction.
- Modify the right hand side of a basic condition, i.e. add or remove a possible value from the set for a categorical attribute, move left/right the upper/lower bound of the interval for a numerical attribute.

For a numerical attribute, which may take a lot of values, the move from a value to the closest possible value may not make any difference. That is the reason why we introduce a range parameter for each numerical attribute. If we denote by v the ordered set of values taken by the attribute in the training set, and $v[i]$ is the current lower bound, moving left that bound will consist in trying to replace $v[i]$ by $v[i - k]$ with, at the beginning of a hill-climbing process, $k = size(v)/2$, we refer to k as the range of the hill-climbing for a given numerical attribute.

The refinement operation is then evaluated by considering all conditions on the resulting aggregate as candidates for a split at the current node of the tree. If no refinement results in an improvement of the metric used to measure split quality (i.e. information gain), then the range of the hill-climbing for numerical attributes present in the conjunction is divided by 2. If this is not possible, i.e. ranges for all attributes in the conjunction are equal to 1, then the hill-climbing process stops and the branch restarts from a newly initialized condition.

At each step of the main loop of the algorithm, the branch on which a step will be performed is chosen through a wheel selection, in which the weight of a branch is determined by the score of the best split ever found by the branch.

A branch is stopped if it has been restarted 3 times without finding a strictly better split, one stopping condition of the algorithm is then when all branches have been stopped. There is also a maximum condition on the number of steps performed globally, on all branches, in the main loop of the algorithm.

The pseudo-code of the main loop of the algorithm is shown in Algorithm 1. The pseudo-code of the algorithm which performs one step of hill-climbing in one branch is shown in Algorithm 2. Algorithm 3 shows a useful subfunction.

5   Experiments and Results

In this section, we first assess the reliability of our algorithm on an artificially generated dataset. Then, we show experimental results on a real-world dataset.

5.1   Experiments on Artificial Data

Description of the Artificial Data. Our artificial dataset consists in a binary classification task, with a main table containing a class attribute. This example is inspired by urban block classification: the main objects are blocks and secondary objects are buildings that make up the blocks. The schema is given in Fig. 2. There are 200 instances of urban blocks in the dataset. The buildings (secondary table) have 4 numerical attributes generated using a random draw

Algorithm 1. Random Restart Hill-Climbing Algorithm

1: **Input:** *functions:* list of aggregation functions, *features:* list of attributes of the secondary table, *train:* labelled training set
2: **Output:** *split:* best complex aggregate found through hill-climbing

3: wheel ← InitializeBranches(functions, features)
4: bestSplits ← []
5: bestScore ← WORST_SCORE_FOR_METRIC
6: **for** i = 1 **to** MAX_ITERATIONS **and** wheel contains at least one branch **do**
7:      branch ← ChooseBranchToGrow(wheel)
8:      hasImproved ← branch.Grow(train)
9:      **if not** hasImproved **then**
10:        **if** branch.split.score ≥ bestScore **then**
11:          **if** branch.split.score > bestScore **then**
12:            bestScore ← branch.split.score
13:            bestSplits ← []
14:          **end if**
15:          bestSplits.Add(branch.split)
16:        **end if**
17:        branch.Reinitialize();
18:      **end if**
19: **end for**
20: split ← PickOneAtRandom(bestSplits)
21: **return** split

from an exponential distribution with means 10000 for area, 1000 for perimeter, 100 for elongation and 10 for miller. We use the exponential distribution because [6,9] observed that the values of attributes of buildings follow this distribution in real-world cases. The number of secondary objects associated to a main object has been determined using a random draw from a geometric distribution (i.e. discrete exponential distribution) with mean 12.

The class attribute of the main table is 'yes' if *avg(miller of buildings such that area ≥ 8000 and elongation < 150) ≥ 7*, and 'no' otherwise, i.e. if the inequality is not satisfied, or if the aggregate is not defined because no such building exists in the block.

Experiments. We first measure the accuracy in 10-fold cross-validation of the 3 algorithms:

- RELAGGS used with the decision tree learner J48 implemented in Weka [5],
- TILDE,
- our algorithm, denoted by RRHC-CA (for Random Restart Hill-Climbing of Complex Aggregates).

The results shown in Table 1 are averaged over 5 runs. We observe that our algorithm learns a model for the dataset very well, the accuracy being close to 100 %, while the two other algorithms are less effective on this learning task. This

---

Algorithm 2. Branch.Grow: Hill-Climbing Algorithm for One Branch

---

1: **Input:** *train:* labelled training set
2: **Output:** *hasImproved:* boolean indicating if the step of the hill-climbing has improved the best split found in the current hill-climbing of the branch

---

3: hasImproved ← false
4: **for all** attr ∈ secondary attributes not present in the current aggregation condition **do**
5:     nextAggregate ← aggregate obtained by adding one randomly initialized condition on attr to the aggregation condition of this.aggregate
6:     spl ← EvaluateAggregate(nextAggregate, train)
7:     hasImproved ← hasImproved **or** UpdateBestSplit(spl)
8: **end for**
9: **for all** attr ∈ secondary attributes already present in the current aggregation condition **do**
10:     nextAggregate ← aggregate obtained by removing the condition on attr present in the aggregation condition of this.aggregate
11:     spl ← EvaluateAggregate(nextAggregate, train)
12:     hasImproved ← hasImproved **or** UpdateBestSplit(spl)
13: **end for**
14: **for all** attr ∈ secondary attributes already present in the current aggregation condition **do**
15:     **for all** move ∈ possible moves on the condition on attr present in the aggregation condition of this.aggregate **do**
16:         nextAggregate ← aggregate obtained by applying move to this.aggregate
17:         spl ← EvaluateAggregate(nextAggregate, train)
18:         hasImproved ← hasImproved **or** UpdateBestSplit(spl)
19:     **end for**
20: **end for**
21: **if** (**not** hasImproved) **and** possible to reduce hill-climbing range **then**
22:     Reduce range of hill-climbing
23:     hasImproved ← true
24: **end if**
25: **return** hasImproved

Algorithm 3. Branch.UpdateBestSplit

1: **Input:** *split:* candidate split
2: **Output:** *hasImproved:* boolean indicating if the split is better in terms of split metric than the best split found so far

---

3: hasImproved ← false
4: **if** spl.score > this.split.score **then**
5:     this.split ← spl
6:     this.aggregate ← nextAggregate
7:     hasImproved ← true
8: **end if**
9: **return** hasImproved

---

block_id	class	building_id	area	perimeter	elongation	miller	block_id
1	yes	1_1	10000	900	80	4	1
2	no	1_2	4500	1000	150	6	1
3	no	1_3	5200	1300	100	12	1
...	...	...	...	...	...	...	...
		2_1	12000	850	90	10	2
		2_2	9000	950	180	8	2
		...	...	...	...	...	...

Urban blocks table   Buildings table

Fig. 2. Schema of the synthetic urban block dataset

Table 1. Accuracy in 10-fold cross-validation on the artificial dataset

Algorithm	Accuracy
RELAGGS + J48	66.3 %
TILDE	71 %
RRHC-CA	97.4 %

is due to the complexity of the aggregation condition, which neither RELAGGS nor TILDE can consider.

Then we observe, in every branch, the behaviour of the hill-climbing processes in terms of evolution of the split quality. The results are shown on Fig. 3. Each curve represents the evolution of one hill-climbing process: it shows the evolution of the metric score of the best split built with a complex aggregate during the process so far with respect to the iteration of the hill-climbing process. We focus on two attributes and two aggregation functions only, including the target branch (*mean*, *miller*), which should produce the final aggregate choice.

We observe that the target branch does outperform the others in terms of score at the end of one hill-climbing process. We make the same observation for the branches we do not show. This assesses the ability of our algorithm to find the appropriate aggregation condition for one (*Function*, *Feature*) aggregation context. Moreover, we observe that, in a given branch, a majority of hill-climbing processes reach the same plateau at the end of the process. Following that observation, we decide to remove a branch from the global wheel selection if 3 hill-climbing processes finish without modification of the best aggregate found by the branch.

Then, we look at the gain in metric score that a refinement operation on the aggregation condition achieves. For this, we focus on hill-climbing processes in the target branch. The plots in Fig. 4 show the evolution of the difference in split metric between the modified aggregate and the original one for different hill-climbing processes. The operations shown consist in adding and removing

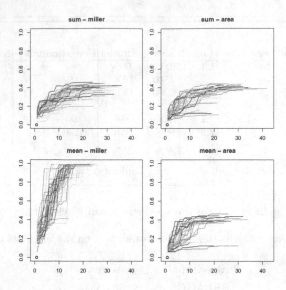

Fig. 3. Evolution of information gain with respect to the number of the iteration in the hill-climbing process in 4 branches (function, attribute to aggregate).

a relevant attribute for the aggregation condition (area) and an irrelevant one (perimeter).

For both addition and removal, we observe clear differences of behaviour. For addition, we observe that once the gain of adding a condition on the area becomes positive, i.e. the condition becomes useful, the modification is chosen by the hill-climbing process and the change is not reverted later in the process since the curve stops, meaning the addition of a condition on the area is never considered again, because one is already present in the aggregation condition. On the other hand, the addition of a condition on the perimeter, which should not be present in the final choice of the algorithm, almost always yields a negative gain, especially when the process is close to its end and converging to the right conjunction of conditions.

For removal, we observe that removing the condition on the area almost always yields a negative gain, especially when the process is converging. This shows why the addition of a condition on the area is never considered again, since the removal of the area is considered a bad refinement choice. On the other hand, the removal of the condition on the perimeter yields a positive gain at the beginning of the process, it is then chosen by the hill-climbing and never considered again because we observed that adding a condition on the perimeter is a bad choice. High variations in gain are due to the difference between random conditions on the attribute when it is added.

To conclude, these experiments show that the addition and removal operations are useful and compensate for a faulty initialization of the aggregation condition, by adding conditions on the relevant attributes and removing conditions

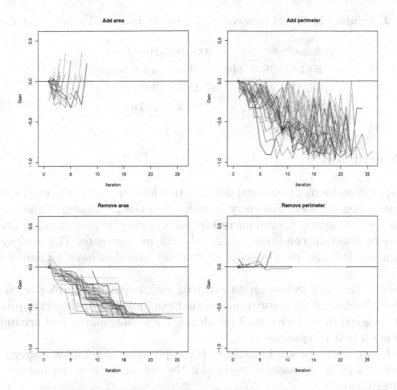

Fig. 4. Gain of addition/removal of conditions on area/perimeter with respect to the number of the iteration in the hill-climbing process.

on the irrelevant ones. Although we do not show them here, the plots for elongation are similar to the plots for area, since it is a relevant attribute for the aggregation condition. Likewise, the plots for miller are similar to the plots for perimeter.

5.2 Experiments on Real-World Data

We evaluate our algorithm on the real-world urban block dataset [10], obtained from previous projects led in collaboration with geographers. The schema of the dataset is as given in Fig. 1. There is a total of 1525 blocks in this dataset, it is a 7-class classification task on which we apply 10-fold cross validation. The accuracy results and runtime are reported in Table 2.

Our algorithm performs best in terms of accuracy. Moreover, we observed the difference between the performance of RRHC-CA and the other algorithms is statistically significant according to paired t-test with 95 % confidence.

Table 2. Accuracy in 10-fold cross-validation on the real-world urban block dataset

Algorithm	Accuracy	Runtime
RELAGGS + J48	91.21 %	a few seconds
TILDE	87.48 %	262 min
RRHC-CA	92.98 %	155 min

6   Conclusion

This paper introduced a relational decision tree learner based on complex aggregate feature construction. For every branch $(Function, Feature)$ of the complex aggregate search space, a random restart hill-climbing is performed to find the appropriate selection condition on the objects to aggregate. The evolution of the branches is made step-by-step, so that all branches have a chance to be improved.

The experimental results on an artificial task requiring complex aggregates show the reliability of our algorithm, and that existing state-of-the-art approaches are not designed to deal with such problems. The results on the real urban block classification task are promising.

For future work, we will investigate the application of complex aggregates to sequence classification tasks, in particular the role of the dimensional attribute in the aggregate.

Acknowledgements. This work is part of the REFRAME project granted by the European Coordinated Research on Long-term Challenges in Information and Communication Sciences & Technologies ERA-Net (CHIST-ERA).

References

1. Assche, A.V., Vens, C., Blockeel, H., Dzeroski, S.: First order random forests: learning relational classifiers with complex aggregates. Mach. Learn. **64**(1–3), 149–182 (2006)
2. Blockeel, H., Dehaspe, L., Ramon, J., Struyf, J., Assche, A.V., Vens, C., Fierens, D.: The ACE Data Mining System, March 2009
3. Blockeel, H., Raedt, L.D.: Top-down induction of first-order logical decision trees. Artif. Intell. **101**(1–2), 285–297 (1998)
4. Getoor, L.: Multi-relational data mining using probabilistic relational models: research summary. In: Proceedings of the First Workshop in Multi-relational Data Mining (2001)
5. Hall, M., Frank, E., Holmes, G., Pfahringer, B., Reutemann, P., Witten, I.H.: The Weka data mining software: an update. SIGKDD Explor. Newslett. **11**(1), 10–18 (2009). http://doi.acm.org/10.1145/1656274.1656278
6. El Jelali, S., Braud, A., Lachiche, N.: Propositionalisation of continuous attributes beyond simple aggregation. In: Riguzzi, F., Železný, F. (eds.) ILP 2012. LNCS, vol. 7842, pp. 32–44. Springer, Heidelberg (2013)

7. Joshi, S., Ramakrishnan, G., Srinivasan, A.: Feature construction using theory-guided sampling and randomised search. In: Železný, F., Lavrač, N. (eds.) ILP 2008. LNCS (LNAI), vol. 5194, pp. 140–157. Springer, Heidelberg (2008). http://dx.doi.org/10.1007/978-3-540-85928-4_14
8. Krogel, M.A., Wrobel, S.: Facets of aggregation approaches to propositionalization. In: Horvath, T., Yamamoto, A. (eds.) Work-in-Progress Track at the Thirteenth International Conference on Inductive Logic Programming (ILP) (2003)
9. Puissant, A., Lachiche, N., Skupinski, G., Braud, A., Perret, J., Mas, A.: Classification et évolution des tissus urbains à partir de données vectorielles. Revue Internationale de Géomatique **21**(4), 513–532 (2011)
10. Ruas, A., Perret, J., Curie, F., Mas, A., Puissant, A., Skupinski, G., Badariotti, D., Weber, C., Gancarski, P., Lachiche, N., Lesbegueries, J., Braud, A.: Conception of a GIS-platform to simulate urban densification basedon the analysis of topographic data. In: Ruas, A. (ed.) Advances in Cartography and GIScience. LNGC, vol. 2, pp. 413–430. Springer, Heidelberg (2011). http://dx.doi.org/10.1007/978-3-642-19214-2_28
11. Vens, C., Ramon, J., Blockeel, H.: Refining aggregate conditions in relational learning. In: Fürnkranz, J., Scheffer, T., Spiliopoulou, M. (eds.) PKDD 2006. LNCS (LNAI), vol. 4213, pp. 383–394. Springer, Heidelberg (2006)

Logical Minimisation of Meta-Rules Within Meta-Interpretive Learning

Andrew Cropper[⊠] and Stephen H. Muggleton

Department of Computing, Imperial College London, London, UK
a.cropper13@imperial.ac.uk

Abstract. Meta-Interpretive Learning (MIL) is an ILP technique which uses higher-order meta-rules to support predicate invention and learning of recursive definitions. In MIL the selection of meta-rules is analogous to the choice of refinement operators in a refinement graph search. The meta-rules determine the structure of permissible rules which in turn defines the hypothesis space. On the other hand, the hypothesis space can be shown to increase rapidly in the number of meta-rules. However, methods for reducing the set of meta-rules have so far not been explored within MIL. In this paper we demonstrate that irreducible, or minimal sets of meta-rules can be found automatically by applying Plotkin's clausal theory reduction algorithm. When this approach is applied to a set of meta-rules consisting of an enumeration of all meta-rules in a given finite hypothesis language we show that in some cases as few as two meta-rules are complete and sufficient for generating all hypotheses. In our experiments we compare the effect of using a minimal set of meta-rules to randomly chosen subsets of the maximal set of meta-rules. In general the minimal set of meta-rules leads to lower runtimes and higher predictive accuracies than larger randomly selected sets of meta-rules.

1 Introduction

In [11] techniques were introduced for predicate invention and learning recursion for regular and context-free grammars based on instantiation of predicate symbols within higher-order definite clauses, or *meta-rules*. The approach was extended to a fragment of dyadic datalog in [10,12] and shown to be applicable to problems involving learning kinship relations, robot strategies and dyadic concepts in the NELL database [4]. More recently MIL was adapted to bias reformulation when learning a hierarchy of dyadic string transformation functions for spreadsheet applications [8]. To illustrate the idea consider the meta-rule below.

Name	Meta-Rule
Chain	$P(x,y) \leftarrow Q(x,z), R(z,y)$

By using meta-rules as part of the proof of an example, an adapted meta-interpreter is used to build up a logic program based on a sequence of meta-substitutions into the given meta-rules which produce rules such as the following

© Springer International Publishing Switzerland 2015
J. Davis and J. Ramon (Eds.): ILP 2014, LNAI 9046, pp. 62–75, 2015.
DOI: 10.1007/978-3-319-23708-4_5

instance of the meta-rule above.

$$\text{aunt}(x, y) \leftarrow \text{sister}(x, z), parent(z, y)$$

1.1 Motivation

In [8] it was shown that within the H_2^2 hypothesis space[1] the number of programs of size n which can be built from p predicate symbols and m meta-rules is $O(m^n p^{3n})$. This result implies that the efficiency of search in MIL can be improved by reducing the number of meta-rules. In this paper we investigate ways in which minimisation of the number of meta-rules can be achieved, without loss of expressivity, by employing logical reduction techniques. This idea is illustrated in Fig. 1. The two meta-rules on the left of Fig. 1 will be shown in Sect. 4.2 to form a minimal set for chained H_2^2. Application of such meta-rules within the meta-interpreter described in Sect. 5 leads to the introduction of clauses involving invented predicates such as those found in the second column of Fig. 1. These definitions with invented predicates allow the search to explore the same space as if it had the meta-rules shown in the third column of Fig. 1 which are found in the resolution closure of the minimal set of two meta-rules. The results in Sect. 4.2 show that in fact meta-rules with aribitrary numbers of literals in the clause body can be imitated in this way by the two minimal meta-rules in Fig. 1.

This paper is organised as follows. Related work is discussed in Sect. 2. In Sect. 3 we describe meta-rules and the use of encapsuation. We then use encapsulation in Sect. 4 to show how sets of meta-rules can be reduced. This is followed in Sect. 5 by a description of the Prolog implementation of the Meta-Interpretretive Learning algorithm used in the experiments. In the Experiments in Sect. 6 we show that reduction of the meta-rules and background knowledge leads to reduced learning times with increased predictive accuracy of the learned theories. In Sect. 7 we conclude the paper.

2   Related Work

Many approaches have been used to improve search within Inductive Logic Programming. These include probabilistic search techniques [16], the use of query packs [2] and the use of special purpose hardware [1]. By contrast this paper explores the possibility of provably correct techniques for speeding up search based on automatically reducing the inductive meta-bias of the learner.

The use of meta-rules to build a logic program is, in some ways[2], analogous to the use of refinement operators in ILP [13,15] to build a definite clause literal-by-literal. As with refinement operators, it seems reasonable to ask about

---

[1] H_j^i consists of definite datalog logic programs with predicates of arity at most i and at most j atoms in the body of each clause.

[2] It should be noted that MIL uses example driven test-incorporation for finding consistent programs as opposed to the generate-and-test approach of clause refinement.

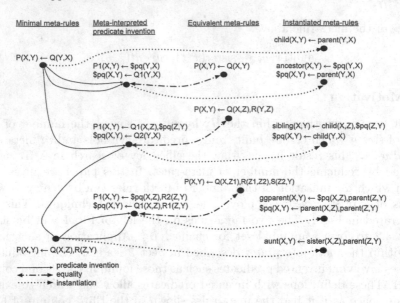

Fig. 1. The relationship between the minimal set of meta-rules, meta-rules via meta-interpreted predicate invention, and instantiated meta-rules.

completeness and irredundancy of any particular set of meta-rules. This question, which has not been addressed in previous papers on MIL, is investigated in Sect. 3. On the other hand, unlike refinement operators, meta-rules can be viewed as a form of declarative bias [6]. By comparison with other forms of declarative bias in ILP, such as modes [9,17] or grammars [5], meta-rules are logical statements. This provides the potential for reasoning about them and manipulating them alongside normal first-order background knowledge. However, in order to do so, we need a method which supports reasoning over a mixture of higher-order and first-order clauses.

3   Meta-Rules and Encapsulation

In this section we introduce a program transformation technique, called *encapsulation*, which supports reasoning over meta-rules by transforming higher-order and first-order datalog statements into first-order definite clauses. We show that a higher-order datalog program has a model whenever the encapsulation of the program has a model. This allows us to show that logical reduction of encapsulated meta-rules and background knowledge can be used as a method of reducing redundancy when searching the hypothesis space. Moreover, by applying logical reduction to a complete set of meta-rules for any fragment of first-order logic we can produce a compact set of meta-rules which is complete by construction. We assume standard logic programming notation [13] throughout this section and Sect. 4.

3.1 Specificational and Derivational Meta-Rules

In Sect. 1 we introduced the example shown below of a meta-rule M and an associated derived clause C.

Meta-Rule M	$P(x,y) \leftarrow Q(x,z), R(z,y)$
Clause C	$aunt(x,y) \leftarrow parent(x,z), sister(z,y)$

M can be viewed as a template for C. In both the meta-rule and the clause the variables x, y, z are universally quantified while in the meta-rule the variables P, Q, R are unbound (i.e. have no quantification). When used in the context of reasoning the unbound variables can have quantifications bound in two separate ways.

Existential quantification. Binding existential quantifiers to the unbound variables in a meta-rule produces a form of *specification* for clauses to be generated in hypotheses, so with this binding we refer to the resulting rules as *specificational meta-rules*. As in formal methods, we assume that if the specificational meta-rule M_S is the specification for clause C then $C \models M_S$.

Universal quantification. Binding universal quantifiers to the unbound variables $\mathcal{V}$ in a meta-rule produces a form appropriate for checking *derivability* for clauses in hypotheses, so with this binding we refer to the resulting rules as *derivational meta-rules*. A clause C is derivable from a derivational meta-rule M_D in the case that there exists a substitution θ with domain $\mathcal{V}$ such that $M_D\theta = C$, implying that $M_D \models C$.

In [10, 12] specificational meta-rules were referred to simply as *meta-rules*. In this paper we distinguish between specificational and derivational meta-rules in order to clarify different forms of logical reasoning used in reduction of logical formulae involving first-order and higher-order statements.

3.2 Encapsulation

In order to support reasoning over specificational and derivational meta-rules using first-order logic we introduce a mechanism called *encapsulation*.

Definition 1 Atomic encapsulation. *Let A be higher-order or first-order atom of the form $P(t_1, .., t_n)$. We say that $enc(A) = m(P, t_1, .., t_n)$ is an encapsulation of A.*

Note that the encapsulation of a higher-order atom is first-order. We now extend atomic encapsulation to logic programs.

Definition 2 Program encapsulation. *The logic program $enc(P)$ is an encapsulation of the higher-order definite datalog program P in the case $enc(P)$ is formed by replacing all atoms A in P by $enc(A)$.*

Case	Meta-rules and clause	Encapsulated clause
a)	$\$p(X,Y) \leftarrow \$q(X,Z), \$r(Z,Y)$	$m(\$p, X, Y)) \leftarrow m(\$q, X, Z)), m(\$r, Z, Y))$
b)	$P(X,Y) \leftarrow Q(X,Y)$	$m(P, X, Y) \leftarrow m(Q, X, Y)$
c)	$aunt(X,Y) \leftarrow parent(X,Z), sister(Z,Y)$	$m(aunt, X, Y)) \leftarrow m(parent, X, Z)), m(sister, Z, Y))$

Fig. 2. Three cases exemplifying encapsulation of clauses: (a) Specificational meta-rule with existentially quantified variables P, Q, R by Skolem constants $\$p, \$q, \$r$, (b) Derivational meta-rule with universally quantified higher-order variables P, Q, R, (c) a first-order clause.

Figure 2 provides examples of the encapsulation of specificational and derivational meta-rules and a first-order clause. The notion of encapsulation can readily be extended to interpretations of programs.

Definition 3 Interpretation encapsulation. *Let I be a higher-order interpretation over the predicate symbols and constants in signature Σ. The encapsulated interpretation $enc(I)$ is formed by replacing each atom A in I by $enc(A)$.*

We now have the following proposition.

Proposition 1 First-order models. *The higher-order datalog definite program P has a model M if and only if $enc(P)$ has the model $enc(M)$.*
Proof. *Follows trivially from the definitions of encapsulated programs and interpretations.*

We now have a method of defining entailment between higher-order datalog definite programs.

Proposition 2 Entailment. *For higher-order datalog definite programs P, Q we have $P \models Q$ if and only if every model $enc(M)$ of $enc(P)$ is also a model of $enc(Q)$.*
Proof. *Follows immediately from Proposition 1.*

4   Logically Reducing Meta-Rules

In [14] Plotkin provides the following definitions as the basis for eliminating logically redundant clauses from a first-order clausal theory.

Definition 4 Clause redundancy. *The clause C is logically redundant in the clausal theory $T \wedge C$ whenever $T \models C$.*

Note that if C is redundant in $T \wedge C$ then T is logically equivalent to $T \wedge C$ since $T \models T \wedge C$ and $T \wedge C \models T$. Next Plotkin defines a reduced clausal theory as follows.

Definition 5 Reduced clausal theory. *Clausal theory T is reduced in the case that it does not contain any redundant clauses.*

Plotkin uses these definitions to define a simple algorithm which given a first-order clausal theory T repeatedly identifies and removes redundant clauses until the resulting clausal theory T' is reduced.

Set of Meta-rules	Encapsulated Meta-rules	Reduced set
P(X,Y) ← Q(X,Y)	m(P,X,Y) ← m(Q,X,Y)	
P(X,Y) ← Q(Y,X)	m(P,X,Y) ← m(Q,Y,X)	m(P,X,Y) ← m(Q,Y,X)
P(X,Y) ← Q(X,Y), R(X,Y)	m(P,X,Y) ← m(Q,X,Y), m(R,X,Y)	
P(X,Y) ← Q(X,Y), R(Y,X)	m(P,X,Y) ← m(Q,X,Y), m(R,Y,X)	
P(X,Y) ← Q(X,Z), R(Y,Z)	m(P,X,Y) ← m(Q,X,Z), m(R,Y,Z)	
P(X,Y) ← Q(X,Z), R(Z,Y)	m(P,X,Y) ← m(Q,X,Z), m(R,Z,Y)	m(P,X,Y) ← m(Q,X,Z), m(R,Z,Y)
P(X,Y) ← Q(Y,X), R(X,Y)	m(P,X,Y) ← m(Q,Y,X), m(R,X,Y)	
P(X,Y) ← Q(Y,X), R(Y,X)	m(P,X,Y) ← m(Q,Y,X), m(R,Y,X)	
P(X,Y) ← Q(Y,Z), R(X,Z)	m(P,X,Y) ← m(Q,Y,Z), m(R,X,Z)	
P(X,Y) ← Q(Y,Z), R(Z,X)	m(P,X,Y) ← m(Q,Y,Z), m(R,Z,X)	
P(X,Y) ← Q(Z,X), R(Y,Z)	m(P,X,Y) ← m(Q,Z,X), m(R,Y,Z)	
P(X,Y) ← Q(Z,X), R(Z,Y)	m(P,X,Y) ← m(Q,Z,X), m(R,Z,Y)	
P(X,Y) ← Q(Z,Y), R(X,Z)	m(P,X,Y) ← m(Q,Z,Y), m(R,X,Z)	
P(X,Y) ← Q(Z,Y), R(Z,X)	m(P,X,Y) ← m(Q,Z,Y), m(R,Z,X)	

Fig. 3. Encapsulation of all 14 distinct specificational chained meta-rules from $H_m^{i=2}$ leading to a reduced set of two.

4.1 Reduction of Pure Dyadic Meta-Rules in $H_2^{i=2}$

We define $H_m^{i=2}$ as the subclass of higher-order dyadic datalog programs with lieterals of arity 2 and clause bodies consisting of at most m atoms. Now we define an associated class of chained meta-rules which will be used in the experiments.

Definition 6 Chained subset of $H_m^{i=2}$. *Let C be a meta-rule in $H_m^{i=2}$. Two literals in C are connected if and only if they share a variable or they are both connected to another literal in C. C is in the chained subset of $H_m^{i=2}$ if and only if each variable in C appears in exactly two literals and a path connects every literal in the body of C to the head of C.*

We generated a set of all chained meta-rules $H_m^{i=2}$ for this class, of which there are 14, and ran Plotkin's clausal theory reduction algorithm on this set. Figure 3 shows the 14 meta-rules together with their encapsulated form and the result of running Plotkin's algorithm which reduces the set to two meta-rules. Since, by construction this set is logically equivalent to the complete set of 14 it can be considered a *universal* set (sufficient to generate all hypotheses) for chained $H_2^{i=2}$. In the following section, we prove the same minimal set of meta-rules are complete for $H_m^{i=2}$.

4.2 Completeness Theorem for $H_m^{i=2}$

We now show that two meta-rules are sufficient to generate all hypotheses in $H_m^{i=2}$.

Definition 7. *A $H_m^{i=2}$ chain rule is a meta-rule of the form $P(U_1, V) \leftarrow T_1(U_1, U_2), \ldots, T_m(U_m, V)$ where $m > 0$, P and each T_i are existentially quantified distinct predicate variables, and V and each U_i are universally quantified distinct term variables.*

We restrict ourselves to *chained* $H_m^{i=2}$. This subclass of H_m^2 is sufficiently rich for the kinship and robot examples in Sect. 6. We now introduce two elementary meta-rules.

Definition 8. *Let $C = P(X, Y) \leftarrow Q(Y, X)$. Then C is the inverse rule.*

Definition 9. *Let $C = P(X, Y) \leftarrow Q(X, Z), R(Z, Y)$. Then C is the $H_2^{i=2}$ chain rule.*

We now show that the inverse rule and the $H_2^{i=2}$ chain rule are sufficient to generate all hypotheses in *chained $H_m^{i=2}$*.

Lemma 1. *Let C be in $H_m^{i=2}$, L a literal in the body of C, and $m > 0$. Then resolving L with the head of the inverse rule gives the resolvent C' with the literal L' where the variables in L' are reversed.*

Proof. Trivial by construction.

Lemma 2. *Let C be the $H_m^{i=2}$ chain rule and $m > 1$. Then resolving C with the $H_2^{i=2}$ chain rule gives the resolvent R where R is the $H_{m+1}^{i=2}$ chain rule.*

Proof. Assume false. This implies that either a resolvent r of the $H_2^{i=2}$ chain rule with an $H_m^{i=2}$ chain rule does not have $m+1$ literals in the body or that the variables in these literals are not fully chained. Let C be the $H_2^{i=2}$ chain rule s.t. $C = P(X, Y) \leftarrow Q(X, Z), R(Z, Y)$ and let D be the $H_m^{i=2}$ chain rule s.t. $D = S(U1, V) \leftarrow T_1(U_1, U_2), T_2(U_2, U_3), \ldots, T_m(U_m, V)$. Resolving the head of C with the first body literal of D gives the unifier $\theta = \{X/U_1, Y/U_2, P/T_1\}$ and the resolvent $r = S(U_1, V) \leftarrow Q(U_1, Z), R(Z, U_2), T_2(U_2, U_3), \ldots, T_m(U_m, V)$. Since there are $m-1$ literals in $T_2(U_2, U_3), \ldots, T_m(U_m, V)$ then the number of literals in the body of r is $(m - 1) + 2 = m + 1$. Thus owing to the assumption it follows that these literals are not fully chained. Since the variables in D form a chain, it follows that the variables in $T_2(U_2, U_3), \ldots, T_m(U_m, V)$ also form a chain, so all variables in r form a chain. This contradicts the assumption and completes the proof.

Lemma 3. *Let C_1 be the inverse rule, C_2 the $H_m^{i=2}$ chain rule, S_m the set of meta-rules in $H_m^{i=2}$ with exactly m literals in the body, and $m > 0$. Then $\{C_1, C_2\} \models R$ for every R in S_m.*

Proof. R can differ from C_2 in only two ways: (1) the order of the variables in a literal and (2) the order of the literals in the body. Case 1. By Lemma 1 we can resolve any literal L in R with C_1 to derive R' with the literal L' such that the variables in L' are reversed. Case 2. Let $C_2 = P \leftarrow \ldots, T_i, T_j, \ldots, T_m$ and $R = P \leftarrow, \ldots, T_j, T_i \ldots, T_m$ where $T_1, \ldots, T_m$ are literals and the predicate symbols for T_i and T_j are existentially quantified variables. Then we can derive R from C_2 with the substitution $\theta = \{T_i/T_j, T_j/T_i\}$. Since these two cases are exhaustive, this completes the proof.

Theorem 1. *Let C_1 be the inverse rule, C_2 the $H_2^{i=2}$ chain rule, S_m the set of meta-rules in $H_m^{i=2}$, and $m > 0$. Then $\{C_1, C_2\} \models R$ for every R in S_m.*

Proof. By induction on m.

1. Suppose $m = 1$. Then $S_1 = \{(P(X,Y) \leftarrow Q(X,Y)), (P(X,Y) \leftarrow Q(Y,X))\}$. By Lemma 1, we can resolve C_1 with C_1' to derive $C_3 = P(X,Y) \leftarrow Q(X,Y)$. The set $\{C_1, C_3\} = S_1$, thus the proposition is true for $m = 1$.
2. Suppose the theorem holds if $k \leq m$. By Lemma 2 we can resolve the $H_2^{i=2}$ chain rule with the $H_m^{i=2}$ chain rule to derive the $H_{m+1}^{i=2}$ chain rule. By the induction hypothesis there is a $H_m^{i=2}$ chain rule, so we can resolve this with C_2 to derive the $H_{m+1}^{i=2}$ chain rule. By Lemma 3, we can derive any R in $H_m^{i=2}$ with exactly m literals, thus completing the proof.

4.3   Representing $H_m^{i=2}$ programs in $H_2^{i=2}$

We now show that $H_2^{i=2}$ is a normal form for $H_m^{i=2}$.

Theorem 2. *Let C be a meta-rule in $H_m^{i=2}$ and $m > 2$. Then there is an equivalent theory in $H_2^{i=2}$.*

Proof. We prove by construction. Let C be of the form $P \leftarrow T_1, \ldots, T_m$. For any literal T_i in the body of C of the form $T(U_{i+1}, U_i)$ introduce a new predicate symbol $\$s_i$, a new clause $(\$s_i(X,Y) \leftarrow T_i(Y,X))$, and then replace $T_i(U_{i+1}, U_i)$ in C with $\$s_i(U_i, U_{i+1})$. Step 2. Introduce new predicates symbols $(\$p_1, \ldots, \$p_{m-2})$ and new clauses $(P(X,Y) \leftarrow T_1(X,Z), \$p_1(Z,Y)), (\$p_1(X,Y) \leftarrow T_2(X,Z), \$p_2(Z,Y)), \ldots, (\$p_{m-2}(X,Y) \leftarrow t_{m-1}(X,Z), t_m(Z,Y))$. Step 3. Remove the original clause C from the theory. You now have an equivalent theory in $H_2^{i=2}$.

This theorem is exemplified below.

Example 1. Let $C = P(U_1, V) \leftarrow T_1(U_1, U_2), T_2(U_3, U_2), T_3(U_3, U_4), T_4(U_4, V)$. Notice that literal T_2 is in non-standard form, i.e. the variables in T_2 are in the order (U_{i+1}, U_i) and not (U_i, U_{i+1}). We now proceed to a theory in $H_2''^{i=2}$ equivalent to C. Step 1. Introduce a new predicate symbol $\$s_2$, a new clause $\$s_2(X,Y) \leftarrow T_3(Y,X)$, and replace T_3 in C with $\$s_2(U_2, U_3)$ so that $C = P(U_1, V) \leftarrow T_1(U_1, U_2), \$s_2(U_2, U_3), T_3(U_3, U_4), T_4(U_4, V)$. Step 2. Introduce new predicate symbols $\$p_1$ and $\$p_2$ and the following new clauses:

$$P(X,Y) \leftarrow T_1(X,Z), \$p_1(Z,Y)$$
$$\$p_1(X,Y) \leftarrow \$s_2(X,Z), \$p_2(Z,Y)$$
$$\$p_2(X,Y) \leftarrow T_3(X,Z), T_4(Z,Y)$$

Step 3. After removing C you are left with the following theory, which is equivalent to C.

$$\$s_2(X,Y) \leftarrow T_2(Y,X)$$
$$P(X,Y) \leftarrow T_1(X,Z), \$p_1(Z,Y)$$
$$\$p_1(X,Y) \leftarrow \$s_2(X,Z), \$p_2(Z,Y)$$
$$\$p_2(X,Y) \leftarrow T_3(X,Z), T_4(Z,Y)$$

5   Implementation

Figure 4 shows the implementation of MetagolD used in the experiments in
Sect. 6. In this implementation the generalised Meta-Interpreter (Fig. 4a) has
a similar form to a standard Prolog meta-interpreter. The differences are as
follows. While Prolog program represents its program by a set of first-order definite clauses, Metagol represents its program by a set of meta-substitutions which
yield first-order clauses when applied to the associated meta-rules. When a Prolog meta-interpreter repreatedly unifies the first atom in the goal with the head
of a clause selected from the clause base it updates the bindings in the answer
set. By contrast, when the meta-interpretrive learner in Fig. 4a unifies *Atom* in
the goal with the head of the meta-rule *Atom :- Body* from the meta-rule base
the *abduce* predicate adds the meta-substitution *MetaSub* to the set of meta-
substitutions comprising the hypothesised program[3]. Figure 4b shows the set of
dyadic meta-rules used within the experiments.

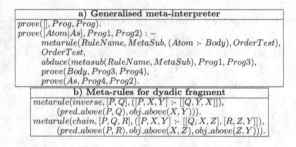

Fig. 4. MetagolD Prolog representation of (a) Generalised meta-interpreter and (b)
Dyadic fragment meta-rules

6   Experiments

These experiments compare the effect of using a minimal set of meta-rules to a
maximum set of meta-rules. We also consider the intermediate cases, i.e. randomly chosen subsets of the maximum set of meta-rules. We define $Metagol_{min}$
and $Metagol_{max}$ as variants of $Metagol_D$ [10,12] which use the minimum and
maximum sets of meta-rules respectively,

6.1   Experimental Hypotheses

The following null hypotheses were tested:

Null Hypothesis 1. $Metagol_{min}$ cannot achieve higher predictive accuracy
than $Metagol_{max}$.

Null Hypothesis 2. $Metagol_{min}$ and $Metagol_{max}$ have the same learning times.

---

[3] The *OrderTest* represents a meta-rule associated constraint which ensures termination, as explained in [10,12].

6.2   Learning Kinship Relations

We used the kinship dataset from [7][4], which contains 12 dyadic relations: *aunt, brother, daughter, father, husband, mother, nephew, niece, sister, son,* and *wife,* and 104 examples.

Experiment 1: Metagol$_{min}$ vs Metagol$_{max}$. We randomly selected n examples, half positive and half negative, of each target kinship relation and performed leave-one-out-cross-validation. All relations excluding the target relation were used as background knowledge. Predictive accuracies and associated learning times were averaged over each relation over 200 trials. Figure 5 shows that Metagol$_{min}$ outperforms Metagol$_{max}$ in predictive accuracy. A McNemar's test at the 0.001 level confirms the significance, rejecting null hypothesis 1. This can be explained by the larger hypothesis space searched by Metagol$_{max}$ and the Blumer bound [3], which says that a larger search space leads to higher predictive error. Figure 5 shows that Metagol$_{min}$ outperforms Metagol$_{max}$ in learning time with Metagol$_{max}$ taking up to seven times longer. Thus null hypothesis 2 is clearly rejected.

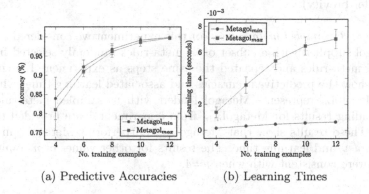

(a) Predictive Accuracies   (b) Learning Times

Fig. 5. Performance of full enumeration (Metagol$_{max}$) and logically reduced (Metagol$_{min}$) sets of metarules when varying number of training examples.

Experiment 2: Sampled background relations. To explore the effect of partial background information we repeated the same steps as in experiment 1 but randomly selected p relations as the background knowledge. Figure 6(a) shows that Metagol$_{max}$ outperforms Metagol$_{min}$ in predictive accuracy when n=4, but when the number of background relations approaches eight Metagol$_{min}$ starts to outperform Metagol$_{max}$. This is because Metagol$_{min}$ has to derive meta-rules (e.g. the identity rule) through predicate invention to replace missing relations and is reaching the logarithmic depth bound on the iterative deepening search.

---

[4] https://archive.ics.uci.edu/ml/datasets/Kinship.

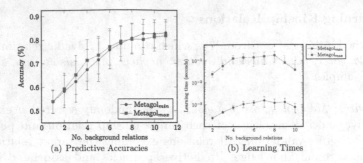

Fig. 6. Performance of full enumeration (Metagol$_{max}$) and logically reduced (Metagol$_{min}$) sets of metarules when varying number of background relations.

A subsequent experiment performed without a logarithmic depth bound confirmed this. Figure 6(b) shows that Metagol$_{max}$ has considerably longer learning time compared to Metagol$_{min}$. We repeated this experiment for $n>4$ but as the number relations approaches six the learning time of Metagol$_{max}$ became prohibitively slow, whereas Metagol$_{min}$ was still able to learn definitions (results omitted for brevity).

Experiment 3: Sampled meta-rules. For this experiment we considered a version of Metagol supplied with a subset of m meta-rules randomly selected from the set of all meta-rules and repeated the same steps as experiment 1. Figure 7(a) and (b) show the predictive accuracies and associated learning times when $n=4$ where Metagol$_m$ represents Metagol loaded with m sampled meta-rules. The corresponding results for Metagol$_{min}$ from experiment 1 are provided for comparison. These results show that Metagol$_{min}$ outperforms Metagol$_m$ in predictive accuracy and learning time. The results for other values of n (omitted for brevity) were consistent with when $n=4$.

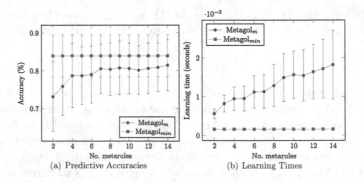

Fig. 7. Peformance when sampling metarules from the full enumeration.

6.3   Learning Robot Strategies

Imagine a robot in a 2-dimensional space which can perform six dyadic actions: *move_left, move_right, move_forwards, move_backwards, grab_ball,* and *drop_ball*. The robot's state is represented as a list with three elements [Robot-Pos,BallPos,HasBall], where RobotPos and BallPos are coordinates and HasBall is a boolean representing whether the robot has the ball. The robot's task is to move the ball to a destination. Suppose we have the following example:

$$move\_ball([0/0, 0/0, false], [2/2, 2/2, false])$$

Here the robot and ball both start at (0,0) and both end at (2,2). Given this example, Metagol$_{min}$ learns the following strategy:

$$s2(X, Y) \leftarrow move\_forwards(X, Z), drop\_ball(Z, Y)$$
$$s2(X, Y) \leftarrow grab\_ball(X, Z), move\_right(Z, Y)$$
$$s3(X, Y) \leftarrow s2(X, Z), s2(Z, Y)$$
$$move\_ball(X, Y) \leftarrow s3(X, Z), s3(Z, Y)$$

Here the robot grabs the ball, moves right, moves forward, drops the ball, and then repeats this process. Metagol$_{min}$ also learns an alternative strategy where the robot performs grab_ball and drop_ball only once. Both are considered equal because both are the same length. If we wanted to prefer solutions which minimise the number of grabs and drops we would need to associate costs with different operations. Now suppose that we exclude move_right from the background knowledge, given the original example, Metagol$_{min}$ learns the strategy:

$$s2(X, Y) \leftarrow move\_backwards(X, Z), drop\_ball(Z, Y)$$
$$s2(X, Y) \leftarrow grab\_ball(X, Z), move\_left(Z, Y)$$
$$s3(X, Y) \leftarrow s2(X, Z), s2(Z, Y)$$
$$s4(X, Y) \leftarrow s3(X, Z), s3(Z, Y)$$
$$move\_ball(X, Y) \leftarrow s4(Y, X)$$

Metagol$_{min}$ found this solution by inverting the high-level action $s4$, thus allowing the actions *move_left, move_backwards,* and *grab_ball* to indicate their respective inverse actions *move_right, move_forwards,* and *drop_ball*. If we also remove move_forwards and drop_ball from the background knowledge, Metagol$_{min}$ learns a strategy which uses only five clauses and three primitives, compared to the original four clause solution which used six primitives. The construction of an inverse plan is familiar to retrograde analysis of positions in chess, in which you go backwards from an end position to work out the moves necessary to get there from a given starting position. We compared Metagol$_{min}$ and Metagol$_{max}$ for learning robot strategies, but the learning time of Metagol$_{max}$ was prohibitively slow in all but trivial circumstances, whereas Metagol$_{min}$ was able to find optimal solutions with acceptable learning times.

7 Conclusion and Future Work

We have shown that just two meta-rules are complete and sufficient for generating all hypotheses in the case of $H_2^{i=2}$. However, this set of reduced meta-rules is insufficient for generating all hypotheses in the case of H_2^2. For example, there is no way to generate any hypothesis containing monadic predicates. Future work should address this.

Our experiments suggest that the minimal set of meta-rules achieves higher predictive accuracies and lower learning times than the maximum set. In Sect. 6.2 we explored the intermediate cases by sampling meta-rules from the maximum set. The results suggested that there was no benefit in using more meta-rules than the minimum set. However, this is not always the case. For example, the dyadic string transformation functions described in [8] require only the chain meta-rule. Thus the optimal set of meta-rules for this task is clearly a subset of the reduced minimal sets described in this paper. One future direction is to investigate learning the optimal set of meta-rules as to minimise the hypothesis space. One idea is to start with a single meta-rule and to invent new meta-rules when required. This would require a MIL equivalent of refinement operators and the need to develop a theory regarding subsumption order between meta-rules. Closely related to this is learning orderings of meta-rules. For example, when learning robot strategies in Sect. 6.3 the chain rule was used more often than the inverse rule. It would be preferable to learn an ordering for the meta-rules to prevent the search from entering unnecessary branches.

7.1 Future Work

Results in this paper have been developed for the case of finding minimal sets of meta-rules for the chained $H_m^{i=2}$ class of definite clause programs. In future work we intend to see whether the same approach can be extended to broader classes of meta-rules.

In addition, the ability to encapsulate background knowledge, as demonstrated in Fig. 2 indicates that it might be possible to minimise the meta-rules together with a given set of background clauses. Preliminary experiments seem to indicate that this is possible, and we would aim to report results on this approach in any extended version of this paper.

Finally, when learning robot strategies in Sect. 6.3 we observed a situation when Metagol learned hypotheses of equal length, but one was preferable to the other. This suggests future work investigating a version of MIL which associates costs with different operations, so that Metagol prefers solutions with lower costs. This would have applications in robot strategy learning and path planning problems.

Acknowledgements. The first author acknowledges the support of the BBSRC and Syngenta in funding his PhD Case studentship. The second author would like to thank the Royal Academy of Engineering and Syngenta for funding his present 5 year Research Chair.

References

1. Muggleton, S.H., Fidjeland, A., Luk, W.: Scalable acceleration of inductive logic programs. In IEEE international conference on field-programmable technology, pp. 252–259. IEEE (2002)
2. Blockeel, H., Dehaspe, L., Demoen, B., Janssens, G., Ramon, J., Vandecasteele, H.: Improving the efficiency of inductive logic programming through the use of query packs. J. Artif. Intell. Res. **16**(1), 135–166 (2002)
3. Blumer, A., Ehrenfeucht, A., Haussler, D., Warmuth, M.K.: Learnability and the Vapnik-Chervonenkis dimension. J. ACM **36**(4), 929–965 (1989)
4. Carlson, A., Betteridge, J., Kisiel, B., Settles, B., Hruschka Jr., E.R., Mitchell, T.M.: Toward an architecture for never-ending language learning. In: Proceedings of the Twenty-Fourth Conference on Artificial Intelligence (AAAI 2010) (2010)
5. Cohen, W.: Grammatically biased learning: learning logic programs using an explicit antecedent description language. Artif. Intell. **68**, 303–366 (1994)
6. De Raedt, L.: Declarative modeling for machine learning and data mining. In: Bshouty, N.H., Stoltz, G., Vayatis, N., Zeugmann, T. (eds.) ALT 2012. LNCS, vol. 7568, pp. 12–12. Springer, Heidelberg (2012)
7. Hinton, G.E.: Learning distributed representations of concepts. Artif. Intell. **40**, 1–12 (1986)
8. Lin, D., Dechter, E., Ellis, K., Tenenbaum, J.B., Muggleton, S.H.: Bias reformulation for one-shot function induction. In: Proceedings of the 23rd European Conference on Artificial Intelligence (ECAI 2014), pp. 525–530. IOS Press, Amsterdam (2014)
9. Muggleton, S.H.: Inverse entailment and progol. New Gener. Comput. **13**, 245–286 (1995)
10. S.H. Muggleton and D. Lin. Meta-interpretive learning of higher-order dyadic datalog: Predicate invention revisited. In: Proceedings of the 23rd International Joint Conference Artificial Intelligence (IJCAI 2013), pp. 1551–1557 (2013)
11. Muggleton, S.H., Lin, D., Pahlavi, N., Tamaddoni-Nezhad, A.: Meta-interpretive learning: application to grammatical inference. Mach. Learn. **94**, 25–49 (2014)
12. Muggleton, S.H., Lin, D., Tamaddoni-Nezhad, A.: Meta-interpretive learning of higher-order dyadic datalog: predicate invention revisited. Mach. Learn. **100**(1), 49–73 (2015)
13. Nienhuys-Cheng, S.-H., de Wolf, R.: Foundations of Inductive Logic Programming. LNCS (LNAI), vol. 1228. Springer, Heidelberg (1997)
14. G.D. Plotkin. Automatic methods of inductive inference. PhD thesis, Edinburgh University, August 1971
15. Shapiro, E.Y.: Algorithmic Program Debugging. MIT Press, Cambridge (1983)
16. A. Srinivasan. A study of two probabilistic methods for searching large spaces with ilp. Technical report PRG-TR-16-00, Oxford University Computing Laboratory, Oxford (2000)
17. Srinivasan, A.: The ALEPH manual, Machine Learning at the Computing Laboratory. Oxford University (2001)

Goal and Plan Recognition via Parse Trees Using Prefix and Infix Probability Computation

Ryosuke Kojima[1]([⊠]) and Taisuke Sato[2]

[1] Graduate School of Information Science and Engineering,
Tokyo Institute of Technology, 2-12-1 Ookayama, Meguro-ku, Tokyo, Japan
`kojima@mi.cs.titech.ac.jp`
[2] AI research center, AIST, 2-3-26 Aomi, Koto-ku, Tokyo 135-0064, Japan
`satou.taisuke@aist.go.jp`

Abstract. We propose new methods for goal and plan recognition based on prefix and infix probability computation in a probabilistic context-free grammar (PCFG) which are both applicable to incomplete data. We define goal recognition as a task of identifying a goal from an action sequence and plan recognition as that of discovering a plan for the goal consisting of goal-subgoal structure respectively. To achieve these tasks, in particular from incomplete data such as sentences in a PCFG that often occurs in applications, we introduce prefix and infix probability computation via parse trees in PCFGs and compute the most likely goal and plan from incomplete data by considering them as prefixes and infixes.

We applied our approach to web session logs taken from the Internet Traffic Archive whose goal and plan recognition is important to improve websites. We tackled the problem of goal recognition from incomplete logs and empirically demonstrated the superiority of our approach compared to other approaches which do not use parsing. We also showed that it is possible to estimate the most likely plans from incomplete logs. All prefix and infix probability computation together with the computation of the most likely goal and plan in this paper is carried out using logic-based modeling language PRISM.

Keywords: Prefix probability · PCFG · Plan recognition · Session log

1 Introduction

Goal and plan recognition have been studied in artificial intelligence as an inverse problem of planning and applied to services that require the identification of users' intentions and plans from their actions such as human-computer collaboration [9], intelligent interfaces [4] and intelligent help systems [5][1].

---

[1] In this paper, we distinguish goal recognition and plan recognition; the former is a task of identifying a goal from actions but the latter means to discover a plan consisting of goal-subgoal structure to achieve the goal.

© Springer International Publishing Switzerland 2015
J. Davis and J. Ramon (Eds.): ILP 2014, LNAI 9046, pp. 76–91, 2015.
DOI: 10.1007/978-3-319-23708-4_6

This paper addresses goal and plan recognition using probabilistic context-Free grammars (PCFGs) which are widely used as models not only in natural language processing but also for analyzing symbolic sequences in general. In a simple PCFG model for goal and plan recognition, a symbolized action sequence made by a user is regarded as a sentence generated by a mixture of PCFGs where each component PCFG describes a plan for a specific goal. We call this model *a mixture of PCFGs*. In this setting, the task of goal recognition is to infer the goal from a sentence as the most likely start symbol of some component PCFG in the mixture whereas plan recognition computes the most likely parse tree representing a plan for the goal.

The problem with this simple model is that sentences generated by a grammar alone receive non-zero probability. So the probability of non-sentences is always zero and therefore it is impossible to extract information contained in incomplete sentences by considering their probabilities and (incomplete) parse trees, though they often occur in real data. To overcome this problem, we propose to generalize the probability and parse trees of sentences to those of incomplete sentences. Consider for example a *prefix* of a sentence, i.e., an initial substring of the sentence. It is one type of incomplete sentence and often observed as data when the observation started but is not yet completed such as a medical record of a patient who is receiving treatment[2].

In plan recognition, our proposal enables us to extract the most likely plan from incomplete data bringing us important information about the user's intention. When applied to a website, for example, where we identify the visitor's plan from his/her actions such as clicking links, the discovered plan would reveal the website structure matches the visitor's intention. However, to our knowledge, no grammatical approach so far has extracted a plan from incomplete sentences, much less the most likely plan. In this paper, we show that it is possible to extract the most likely plan from incomplete sentences.

Turning to goal recognition, it is surely possible to directly estimate the goal from actions by feature-based methods such as logistic regression, support vector machines and so on but they are unable to make use of structural information represented by a plan behind the action sequence. We experimentally demonstrate the importance of structural information contained in a plan and compare our method to these feature-based methods in web session log analysis we describe next.

In our web session log analysis, action sequences recorded in session logs are basically regarded as complete sentences in a mixture of PCFGs. We however consider three types of *incomplete data situation*. The first one is the *online situation* where what is desired is navigating visitors to a target web page during their web surfing, say, by displaying affiliate links appropriate to their goals. The second one is *unachieved visitors* who quit the website for some reason before their purpose is achieved. In these situations, their goal is not achieved, and

---

[2] The probability of a prefix in a PCFG is defined to be the sum of probabilities of infinitely many sentences extending it and computed by solving a set of linear equations derived from the CFG [8]. Also there is prefix probability computation based on probabilistic Earley parsing [15].

hence their action sequences should be treated as incomplete sentences. The last type is the *cross-site situation* where a user visits several websites, which is a very likely situation in real web surfing. In this situation an action sequence observed at one website is only part of the whole action sequence. Consequently an action sequence recorded at one website should be considered as an incomplete sentence.

Notice that the first and second situations yield prefixes whereas the last one causes the lack of the beginning part in addition to the ending part (of sentences). This type of incomplete sentence is called *infix*. In this paper, we primarily deal with the first and second situations and touch the last situation. Our analysis is uniformly applicable to all situations made by visitors and can let appropriate advertisements pop up timely on a display during web surfing by detecting visitors' purposes and plans from action sequences recorded in web session logs.

To implement our approach, we use the logic-based modeling language PRISM [12,13] which is a probabilistic extension of Prolog for probabilistic modeling. It supports general and highly efficient probability computation and parameter learning for a wide variety of probabilistic models including PCFGs. For example a PRISM program for PCFGs can perform probability computation in the same time complexity as the Inside-Outside algorithm, a specialized probability computation algorithm for PCFGs. In PRISM, probability is computed by dynamic programming applied to *acyclic explanation graph*s which are an internal data structure encoding all (but finitely many) proof trees. In our previous work [14], we extended them and introduced an efficient mechanism of prefix probability computation based on *cyclic explanation graph*s encoding infinitely many parse trees. In this paper, we further add to PRISM infix probability computation and the associated Viterbi inference which computes the most likely partial parse tree using cyclic explanation graphs obtained from parsing.

In the following, we first review PRISM focusing on prefix probability computation. We next apply prefix probability computation via parse trees to web session log analysis and conduct an experiment on visitors' goals and plans using real data sets. Then infix probability computation via parse trees is briefly described and followed by conclusion.

2 Reviewing PRISM

The prefix and infix probability computation via parse trees proposed in this paper is carried out in PRISM using explanation graphs. So in this section, we quickly review probability computation in PRISM while focusing on prefix probability computation by cyclic explanation graphs.

PRISM is a probabilistic extension of Prolog and provides a basic built-in predicate of the form $\mathtt{msw}(i, v)$ for representing a probabilistic choice used in probabilistic modeling. It is called "multi-valued random switch" ($\mathtt{msw}$ for short) and used to denote a simple probabilistic event $X_i = v$ where X_i is a discrete random variable and v is its realized value. Let $V_i = \{v_1, \cdots, v_{|V_i|}\}$ be the set of possible outcomes of X_i. i is called the *switch name* of $\mathtt{msw}(i, v)$.

To represent the distribution $P(X_i = v)$ $(v \in V_i)$, we introduce a set $\{\mathtt{msw}(i, v) \mid v \in V_i\}$ of mutually exclusive $\mathtt{msw}$ atoms and give them a joint distribution such that $P(\mathtt{msw}(i, v)) = \theta_{i,v} = P(X_i = v)$ $(v \in V_i)$ where $\sum_{v \in V_i} \theta_{i,v} = 1$. $\{\theta_{i,v}\}$ are called *parameters* and the set of all parameters Θ appearing in a program is manually specified by the user or automatically learned from data.

Suppose a positive program DB is given which is a Prolog program containing $\mathtt{msw}$ atoms. We define a *basic distribution* (probability measure) $P_{\mathtt{msw}}(\cdot \mid \Theta)$ as the product of distributions for the $\mathtt{msws}$ appearing in DB. Then it is proved that the basic distribution can uniquely be extended by way of the least model semantics in logic programming to a σ-additive probability measure $P_{DB}(\cdot \mid \Theta)$ over possible Herbrand interpretations of DB. It is the denotation of DB in the *distribution semantics* [11,12] that is a standard semantics for probabilistic logic programming. In the following, we omit Θ when the context is clear.

Semantically PRISM is just one of many possible implementations of the distribution semantics that realizes efficient probability computation by adding two assumptions: *independence* and *exclusiveness* . Let G be a non-$\mathtt{msw}$ atom which is ground. $P_{DB}(G)$, the probability of G defined by the program DB, can be naively computed as follows. First reduce the top-goal G using Prolog's exhaustive top-down proof search to a propositional DNF(disjunctive normal form) formula $\mathrm{expl}_0(G) = e_1 \vee e_2 \vee \cdots \vee e_k$ where $e_i (1 \leq i \leq k)$ is a conjunction of atoms $\mathtt{msw}_1 \wedge \cdots \wedge \mathtt{msw}_n$ such that $e_i, DB \vdash G$[3]. Each e_i is called an *explanation* for G. Then assuming the

Independence condition ($\mathtt{msw}$ atoms in an explanation are independent):

$$P_{DB}(\mathtt{msw} \wedge \mathtt{msw}') = P_{DB}(\mathtt{msw})P_{DB}(\mathtt{msw}')$$

Exclusive condition (explanations are exclusive):

$$P_{DB}(e_i \wedge e_j) = 0 \quad \text{if} \quad i \neq j$$

we compute $P_{DB}(G)$ as

$$P_{DB}(G) = P_{DB}(e_1) + \cdots + P_{DB}(e_k)$$
$$P_{DB}(e_i) = P_{DB}(\mathtt{msw}_1) \cdots P_{DB}(\mathtt{msw}_n) \quad \text{for} \quad e_i = \mathtt{msw}_1 \wedge \cdots \wedge \mathtt{msw}_n$$

Recall that $\mathtt{msws}$ with different switch names are independent by construction of $P_{\mathtt{msw}}(\cdot \mid \Theta)$. We further assume that $\mathtt{msw}$ atoms with the same switch name are *iid* (independent and identically distributed). Fortunately this assumption can be automatically satisfied.

Contrastingly the exclusiveness condition cannot be automatically satisfied. It needs to be satisfied by the user, for example, by writing a program so that it generates an output solely as a sequence of probabilistic choices made by $\mathtt{msw}$

---

[3] $\mathrm{expl}_0(G)$ is equivalent to G in view of the distribution semantics. When convenient, we treat $\mathrm{expl}_0(G)$ as a bag $\{e_1, e_2, \cdots, e_k\}$ of explanations.

atoms (modulo auxiliary non-probabilistic computation). Although most generative models including BNs, HMMs and PCFGs are naturally written in this style, there are models which are not [3]. Relating to this, observe that *Viterbi explanation*, i.e., the most likely explanation e^* for G, is computed similarly to $P_{DB}(G)$ just by replacing sum with argmax: $e^* \overset{\text{def}}{=} \underset{e \in \text{expl}_0(G)}{\text{argmax}} \, P_{DB}(e)$.

So far our computation is naive. Since there can be exponentially many explanations, naive computation would lead to exponential time computation. PRISM avoids this by adopting *tabled search* in the exhaustive search for all explanations for the top-goal G and applying dynamic programming to probability computation. By tabling, a goal which is once called and proved is stored (tabled) in memory with its answer substitutions and later calls to the same goal return with a stored answer substitution without processing further. Tabling is important to probability computation because tabled goals factor out common sub-conjunctions in $\text{expl}_0(G)$, which results in sharing probability computation for common sub-conjunctions, thereby realizing dynamic programming which gives exponentially faster probability computation compared to naive computation.

As a result of exhaustive tabled search for all explanations for G, PRISM yields a set of propositional formulas called *defining formulas* of the form $H \Leftrightarrow B_1 \vee \cdots \vee B_h$ for every tabled goal H that directly or indirectly calls msws. We call the heads of defining formulas *defined goals*. Each $B_i (1 \leq i \leq h)$ is recursively composed of a conjunction $C_1 \wedge \cdots \wedge C_m \wedge \text{msw}_1 \wedge \cdots \wedge \text{msw}_n (0 \leq m, n)$ of defined goals $\{C_1, \cdots, C_m\}$ and msw atoms $\{\text{msw}_1, \cdots, \text{msw}_n\}$. We introduce a binary relation $H \succ C$ over defined goals such that $H \succ C$ holds if H is the head of some defining formula and C occurs in the body. We denote by $\text{expl}(G)$ the whole set of defining formulas and call $\text{expl}(G)$ the *explanation graph* for G as in the non-tabled case. When "$\succ$" is acyclic, we call an explanation graph *acyclic* and extend "$\succ$" to a partial ordering over the defined goals.

Once $\text{expl}(G)$ is obtained as an acyclic explanation graph, since defined goals are layered by the "$\succ$" relation, defining formulas in the bottom layer (minimal elements) have only msws in their bodies whose probabilities are known (declared in the program), so we can compute probabilities by a sum-product operation for all defined goals from the bottom layer upward in a dynamic programming manner in time linear in the number of atoms appearing in $\text{expl}(G)$.

Compared to naive computation, the use of dynamic programming on $\text{expl}(G)$ can reduce time complexity for probability computation from exponential time to polynomial time. For example PRISM's probability computation for HMMs takes $O(L)$ time for a given sequence with length L and coincides with the standard forward-backward algorithm for HMMs. Likewise PRISM's sentence probability computation for PCFGs takes $O(L^3)$ time for a given sentence with length L and coincides with inside probability computation for PCFGs.

Viterbi inference that computes the Viterbi explanation and its probability is similarly performed on $\text{expl}(G)$ in a bottom-up manner like probability computation stated above. The only difference is that we use argmax instead of sum. In what follows, we look into the detail of how the Viterbi explanation is computed.

Let H be a defined goal and $H \Leftrightarrow B_1 \vee \cdots \vee B_h$ the defining formula for H in expl(G). Write $B_i = C_1 \wedge \cdots \wedge C_m \wedge \mathtt{msw}_1 \wedge \cdots \wedge \mathtt{msw}_n (0 \leq m, n)$ $(1 \leq i \leq h)$ and suppose recursively that the Viterbi explanation $e^*_{C_j} (1 \leq j \leq m)$ has already been calculated for each defined goal in C_j in B_i. Then the Viterbi explanation $e^*_{B_i}$ for B_i and Viterbi explanation e^*_H are respectively computed by

$$e^*_{B_i} = e^*_{C_1} \wedge \cdots \wedge e^*_{C_m} \wedge \mathtt{msw}_1 \wedge \cdots \wedge \mathtt{msw}_n$$
$$e^*_H = \underset{B_i}{\arg\max} \, P_{DB}(e^*_{B_i} \mid \Theta) \tag{1}$$
$$\text{where } P_{DB}(e^*_{B_i} \mid \Theta) = P_{DB}(e^*_{C_1}) \cdots P_{DB}(e^*_{C_m}) \theta_{i_1, v_1} \cdots \theta_{i_n, v_n}$$

Here θ_{i_1, v_1} is a parameter associated with $\mathtt{msw}_1$ and so on. In this way, the Viterbi explanation for the top-goal G is computed in a bottom-up manner by scanning expl(G) once in time linear in the size of expl(G) in an acyclic explanation graph.

3   Prefix Probability Computation

In this section, we examine prefix computation for PCFGs in PRISM. A PCFG G_Φ is a CFG G augmented with a parameter set $\Phi = \bigcup_{N^i \in \mathbf{N}} \{\phi_r\}_{N^i}$ where $\mathbf{N}$ is a set of nonterminals, N^1 a start symbol and $\{\phi_r\}_{N^i}$ the set of parameters associated with rules $\{r \mid r = N^i \to \zeta\}$ for a nonterminal N^i where ζ is a sequence of nonterminal and terminal symbols. We assume that the ϕ_r's satisfy $0 < \phi_r < 1$ and $\sum_{\zeta : N^i \to \zeta} \phi_{N^i \to \zeta} = 1$.

There are already algorithms to compute prefix probabilities in PCFGs [8, 15]. We here briefly describe prefix probability computation based on explanation graphs in PRISM [14]. As previously stated, a *prefix* $\mathbf{v}$ is an initial substring of a sentence and the prefix probability $P_{\text{pre}}^{N^1}(\mathbf{v})$ of $\mathbf{v}$ is an infinite sum of probabilities of sentences extending $\mathbf{v}$:

$$P_{\text{pre}}^{N^1}(\mathbf{v}) = \sum_{\mathbf{w}} P_G(\mathbf{vw})$$

where $\mathbf{w}$ ranges over strings such that $\mathbf{vw}$ is a sentence in G. Prefix probabilities are computed in PRISM by way of cyclic explanation graphs. We sketch our prefix probability computation following [14]. We use a PCFG $G_0 = \{$ s $\to$ s s : 0.4, s $\to$ a : 0.3, s $\to$ b : 0.3 $\}$ where "s" is a start symbol and "a" and "b" are terminals and consider the computation of the prefix probability $P_{\text{pre}}^s(\mathtt{a})$ of prefix a. To compute $P_{\text{pre}}^s(\mathtt{a})$, we first parse "a" as a prefix by the PRISM program DB_0 in Fig. 1. As can be seen from the comments, it runs exactly like a standard top-down CFG parser except *pseudo success* at line (6). *pseudo success* means an immediate return with success on the consumption of the input prefix L1 ignoring the remaining nonterminals in R at line (2)[4].

---

[4] This is justified because we assume the consistency of PCFGs [16] that implies the probability of remaining nonterminals in R yielding some terminal sequences is 1.

```
values(s,[[s,s],[a],[b]]).
:- set_sw(s,[0.4,0.3,0.3]).

pre_pcfg(L):- pre_pcfg([s],L,[]).          % (1) L is a prefix
pre_pcfg([A|R],L0,L2):-                     % (2) L0 is ground when called
  ( get_values(A,_) -> msw(A,RHS),          % (3) if A is a nonterminal
      pre_pcfg(RHS,L0,L1)                    % (4) select rule A->RHS
  ; L0=[A|L1] ),                            % (5) else consume A in L0
  ( L1=[] -> L2=[]                          % (6) (pseudo) success
  ; pre_pcfg(R,L1,L2) ).                    % (7) recursion
pre_pcfg([],L1,L1).                          % (8) termination
```

Fig. 1. Prefix parser DB_0

```
pre_pcfg([a]) <=> pre_pcfg([s],[a],[])      : P(pre_pcfg([a])) = X = Y
pre_pcfg([s],[a],[])                         : P(pre_pcfg([s],[a],[])) = Y
  <=> pre_pcfg([s,s],[a],[]) & msw(s,[s,s])      = Z · θ_{s→ss} + W · θ_{s→a}
   v pre_pcfg([a],[a],[]) & msw(s,[a])
pre_pcfg([s,s],[a],[])                       : P(pre_pcfg([s,s],[a],[])) = Z
  <=> pre_pcfg([a],[a],[]) & msw(s,[a])          = W · θ_{s→a} + Z · θ_{s→ss}
   v pre_pcfg([s,s],[a],[]) & msw(s,[s,s])
pre_pcfg([a],[a],[])                         : P(pre_pcfg([a],[a],[])) = W = 1
```

Fig. 2. Explanation graph for prefix "a" (left) and associated probability equations (right)

By running a command ?-probf(pre_pcfg([a])) in PRISM, we obtain an explanation graph in Fig. 2 (left) for pre_pcfg([a])[5]. Note a cycle exists in the explanation graph; pre_pcfg([s,s],[a],[]) calls itself in the third defining formula. Since this is a small example, its explanation graph has only self-loops. In general however, an explanation graph for prefix parsing has larger cycles as well as self-loops, and we call this type of explanation graphs *cyclic explanation graphs* [14].

Then we convert the defining formulas to a set of probability equations about X, Y, Z and W as shown in Fig. 2 (right). We use the assumptions in PRISM that goals are independent ($P(A \wedge B) = P(A)P(B)$) and disjunctions are exclusive ($P(A \vee B) = P(A) + P(B)$). By solving them using parameter values $\theta_{s→ss} = 0.4$ and $\theta_{s→a} = 0.3$ set by :-set_sw(s,[0.4,0.3,0.3]) in the program DB_0, we finally obtain X = Y = Z = 0.5[6]. So we have $P_{pre}^s(a) = X = 0.5$. In general, a set of probability equations generated from a prefix in a PCFG using DB_0 is always linear and solvable by matrix operation [14].

---

[5] probf/1 is a PRISM's built-in predicate and displays an explanation graph.
[6] W = 1 because pre_pcfg([a],[a],[]) is logically proved without involving msws.

We next describe an extension of the Viterbi inference of PRISM to cyclic explanation graphs. The most likely explanation and its probability for cyclic explanation graphs is defined as usual as $e^* \overset{\text{def}}{=} \underset{e \in \text{expl}_0(G)}{\text{argmax}} P_{DB}(e)$ where $\text{expl}_0(G)$ is possibly an infinite set of explanations represented by a cyclic explanation graph. For example, the set of explanations represented by Fig. 2 (left) is $\text{expl}_0(\texttt{pre\_pcfg([s,s],[a],[])}) = \{ \texttt{msw(s,[a])}, \texttt{msw(s,[s,s])} \wedge \texttt{msw(s,[a])}, \texttt{msw(s,[s,s])} \wedge \texttt{msw(s,[s,s])} \wedge \texttt{msw(s,[a])}, \cdots \}$ where the repetition of $\texttt{msw(s,[s,s])}$ is produced by the cycle. Note that although there are infinitely many explanations, the most likely explanation is $\texttt{msw(s,[a])}$ since the product of probabilities is monotonically decreasing w.r.t. the number of occurrences of $\texttt{msw(s,[s,s])}$ $(0 < P_{DB}(\texttt{msw(s,[s,s])})) < 1)$ in an explanation.

The Viterbi algorithm in Eq. (1) for acyclic explanation graphs is no longer applicable to cyclic graphs as it wouldn't stop if applied to them. So we generalize it for cyclic explanation graphs using a shortest path algorithm such as Dijkstra's algorithm and the Bellman-Ford algorithm [6]. In our implementation, we adopted the Bellman-Ford algorithm since it neither requires additional data structure nor memory by reusing the space for the Viterbi algorithm.

4 Action Sequences as Incomplete Sentences in a PCFG

From here on, we tackle the problem of identifying the purposes or goals of visitors who visit a website from their session logs. We first abstract a visitor's session log into a sequence of five basic actions: up, down, sibling, reload and move. The first two, up and down, state that the visitor moves respectively to a page in the parent directory or a subdirectory in the site's directory structure. An action sibling says that the visitor moves to a page in a subdirectory of the parent directory. An action reload means that the visitor requests the same page. An action move categorizes remaining miscellaneous actions. Moving between web pages is expressed by a sequence of basic actions. For example moving from /top/index.html to /top/child/a.html is a down action.

We consider an action sequence generated by a visitor who has achieved the intended goal as a complete sentence in a PCFG. We parse it using rules as in Fig. 3 (left) and obtain a parse tree as illustrated there (right). The CFG rules (left) describe possible goal-subgoal structures behind visitors' action sequences.

Since diverse visitors visit a website with diverse goals in mind, we capture their action sequences $\mathbf{w}$ in terms of a mixture of PCFGs $P(\mathbf{w} \mid N^1) = \sum_A P^A(\mathbf{w} \mid A)P(A \mid N^1)$ where $P^A(\mathbf{w} \mid A)$ is the probability of $\mathbf{w}$ being generated by a visitor whose goal is represented by a nonterminal A and $P(A \mid N^1)$ is the probability of A being derived from the start symbol N^1 respectively. We call such A a *goal-nonterminal* and assume that there is a unique rule $N^1 \rightarrow A$ for each goal-nonterminal A with a parameter $\theta_{N^1 \rightarrow A} = P(A \mid N^1)$.

Finally to make it possible to estimate visitor goals from incomplete sequences, we replace a sentence probability $P^A(\mathbf{w} \mid A)$ in a mixture of PCFGs

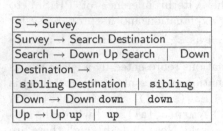

S → Survey
Survey → Search Destination
Search → Down Up Search | Down
Destination →
sibling Destination | sibling
Down → Down down | down
Up → Up up | up

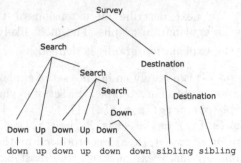

Fig. 3. Example of CFG rules (left) and a parse tree using them (right)

$P(\mathbf{w} \mid N^1) = \sum_A P^A(\mathbf{w} \mid A)P(A \mid N^1)$ by a prefix probability $P_{\mathrm{pre}}^A(\mathbf{w} \mid A)$. We call this method the *prefix method*.

Suppose a prefix $\mathbf{w}_k$ with length k is given as an action sequence. We estimate the most likely goal-nonterminal A^* for $\mathbf{w}_k$ by

$$A^* = \operatorname*{argmax}_A P_{\mathrm{pre}}^A(\mathbf{w}_k)P(A \mid N^1) \tag{2}$$

where A ranges over possible goal-nonterminals. $P_{\mathrm{pre}}^A(\mathbf{w}_k)$ is computed just like $P_{\mathrm{pre}}^{N^1}(\mathbf{w})$ in the previous section.

5   Comparative Experiment

In this section, we empirically evaluate our prefix method and compare it to existing methods: the PCFG method and logistic regression. The PCFG method naively uses a PCFG. It applies a mixture of PCFGs to action sequences $\mathbf{w}_k$ by assuming that every sequence is a sentence. The most likely goal-nonterminal is estimated by substituting the $P_{\mathrm{pre}}^A(\mathbf{w}_k)$ in the Eq. (2) by $P^A(\mathbf{w}_k)$.

We also compare the prefix method with logistic regression which is a popular discriminative model that does not assume any structure behind data unlike the prefix and PCFG methods. For a fixed length k, the most likely visitor goal is estimated from $\mathbf{w}_k$ considered as a feature vector where features are the five basic visitor actions introduced in Sect. 4.

5.1   Data Sets and the Universal Session Grammar

We first prepare three data sets of action sequences by preprocessing the web server logs of U of S (University of Saskatchewan), ClarkNet and NASA [1] in the Internet Traffic Archive [7]. We consider, solely for convenience, action sequences with length greater than 20 as sentences and exclude those with length greater than 30 as the computation of the latter is too costly. In this way we prepared

Table 1. Result of clustering

Cluster (goal-nonterminal)	Features and major action
Survey	up/down moves in the hierarchy of a website
News	up/down moves in the hierarchy of a website + reload the same page
Survey(SpecificAreas)	access to the same layer
News(SpecificAreas)	access to the same layer + reload the same page
Other	others

Table 2. Part of the *universal session grammar*

S → Survey
Survey → InitialSearch Destination EndSearch | Destination EndSearch
Destination→UpDownSearch Search | UpDownSearch
Search→InternalSearch Destination | InternalSearch

three data sets of action sequences referred to here as U of S, ClarkNet and NASA, each containing 652, 4523 and 2014 action sequences respectively.

We next specify a CFG to build a mixture of PCFGs which is applied to the data sets. To do so in turn requires to determine the number of goal-nonterminals. In other words, we have to decide how many goals or intentions visitors have when visiting a website. So we performed clustering on action sequences assuming that one cluster corresponds to one goal, i.e., the number of clusters gives the number of goal-nonterminals. We used a mixture of PCFGs again for clustering[7]. As a result of clustering, we got five clusters which are listed in Table 1.

Finally we manually expand the small CFG used for clustering into a large CFG called the *universal session grammar* that has five goal-nonterminals corresponding to five visitor clusters in Table 1. Some of the rules concerning Survey are listed in Table 2. The universal session grammar contains 102 rules and 32 nonterminals and reflects our observation that visitors have different action patterns between initial, middle and final parts of a session.

5.2 Evaluation of the Prefix Method

We apply the prefix method to the task of estimating the visitors' goals from prefixes of action sequences and record the estimation accuracy while varying

---

[7] Clustering was done by PRISM. We used a small CFG for clustering, containing 30 rules and 12 nonterminals, because clustering by a mixture of large PCFGs tends to suffer from very high memory usage. To build this grammar, we merged similar symbols such as InternalSearch and Search in the universal session grammar shown in Table 2.

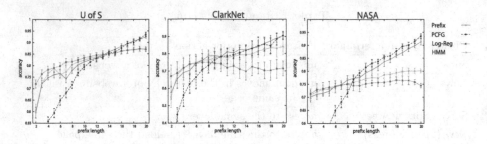

Fig. 4. Accuracy for U of S, ClarkNet and NASA

prefix length. We also apply a mixture of hidden Markov models (HMMs) as a reference method.

To prepare a teacher data set to measure accuracy, we need to label each action sequence by the visitor's true intention or goal, which is however practically impossible. As a substitution, we define a *correct top-goal* for an action sequence in a data set to be the most likely goal-nonterminal for the sequence estimated by a mixture of PCFGs with the universal session grammar whose parameters are learned by the EM algorithm from the data set. This strategy seems to work as long as the universal session grammar is reasonably constructed.

In the experiment[8], accuracy is measured by five-fold cross-validation for each prefix length k ($2 \leq k \leq 20$). After parameter learning by a training data set, prefixes with length k are cut out from action sequences in the test set and their most likely goal-nonterminals are estimated and compared against correct top-goals labeling them. Figure 4 shows accuracy for each k with standard deviation.

Here Prefix denotes the prefix method, PCFG the PCFG method[9] and Log-Reg logistic regression respectively. We also add HMM for comparison which uses a mixture of HMMs instead of a mixture of PCFGs[10,11].

Figure 4 clearly demonstrates that the prefix and PCFG methods outperform logistic regression and HMM when prefix is long. Actually all differences at prefix length $k = 20$ in the graph are statistically significant and confirmed by t-test at 0.05 significance level. Also we can observe, as prefix gets shorter, that the

---

[8] It is conducted on a PC with Core i7 Quad 2.67 GHz, OpenSUSE 11.4 and 72 GB main memory.

[9] We applied a PCFG to prefixes by pretending them to be sentences. In this experiment, we found that the universal session grammar fails to parse at most two sequences for each data set, so we can ignore these sequences.

[10] We used a left-to-right HMM where the number of states is varied from 2 to 8. In Fig. 4, only the highest accuracy is plotted for each k. Since logistic regression only accepts fixed length data, we prepare 19 logistic regression models, one for each length k ($2 \leq k \leq 20$).

[11] We used PRISM to implement a mixture of HMMs and that of PCFG and also to compute prefix probability. For the implementation of logistic regression we used the 'nnet' package of R.

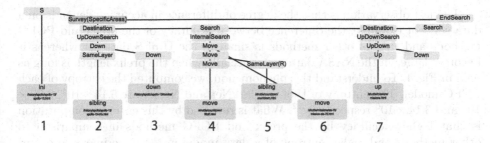

Fig. 5. a prefix parse tree for a action sequence in NASA data

PCFG method rapidly deteriorates though the prefix method keeps fairly good performance comparable to logistic regression and HMM.

At this point we would like to emphasize that our approach can produce a most-likely plan for the estimated goal by the Viterbi algorithm which runs on cyclic explanation graphs. We show an example of estimated plan in Fig. 5. In this figure, the purple node is the estimated goal, the green internal nodes are subgoals and the red leaf nodes stand for actions and web pages accessed by the visitor. Parse trees like this visualize the visitor's plan and help a web manager improve the website. For example in this case, the actions taken from No. 4 to No. 6 are recognized as "Search" and hence we might be able to help the visitor's search by adding a link from the page of No. 3 to that of No. 7.

5.3 Discussion

In the previous section, we experimentally compared the prefix method to existing methods using three probabilistic models: PCFG, logistic regression and HMM. Here we have a closer look at the results of the experiment.

The first observation is that the PCFG method shows poor accuracy when the prefix length is less than 10. This is thought to be caused by a mismatch between the model that assumes the observed data is complete and the incomplete data given as input.

The second one is that the accuracy of the prefix method is higher than or equal to that of logistic regression, a standard discriminative model, when the prefix length is long. Nonetheless, when the length is short, logistic regression outperforms the grammatical methods. This is interpreted as follows. First our grammatical methods require to correctly identify the most-likely parse tree for the top-goal but the identification becomes quite difficult for short sequences as they give little information on correct parse trees. So mis-identification easily occurs which leads to poor performance.

The third one is a notable difference in accuracy between our method and the HMM method. It might be ascribed to the fact that we used the universal session grammar to decide correct answers in the experiment. The use of it as a criterion for accuracy causes a substantial disadvantage to HMM which is a special case of PCFG and not as expressive as the universal session grammar.

The last observation is that the degree of difference in accuracy depends on a data set. For example, the difference between accuracy of the prefix and PCFG methods and that of other methods is small in the U of S data set whereas it becomes larger in the NASA data set, especially when the prefix length is long as seen in Fig. 4. To understand this phenomenon, we computed the entropy of each PCFG model. The entropy of U of S, ClarkNet and NASA are 5.14×10^4, 2.77×10^5 and 3.14×10^6 respectively[12]. What is revealed by this entropy computation is that higher accuracy by the prefix and PCFG methods in comparison to other methods and higher entropy of a data model in the experiment co-occur. We do not think this co-occurrence is accidental. First of all, the entropy is an indicator of uncertainty of a probability distribution and in PCFGs, it represents uncertainty of parse trees. Hence when the data model is simple and the entropy is low, the identification of a correct goal is easy and simple methods such as HMM and logistic regression are comparable to the prefix and PCFG methods. However when the entropy is high as in the case of the NASA data set, these simple approaches fail to disambiguate the correct parse tree and the prefix and PCFG methods that exploit structural information in the input data outperform them, particularly when the data is long.

6 Infix Probability Computation: Beyond Prefix Probability Computation

Up until now we have only considered prefixes that describe session logs under online situation or unachieved visitors. When we consider the cross-site situation however, infix needs to be introduced. Compared to prefix probability computation, infix probability computation is much harder and early attempts put some restrictions on it. However Nederhof and Satta recently proposed a completely general method to compute infix probability that solves a set of non-linear equations [10]. One thing to recall is that their method·is purely numerical and yields no parse trees for prefixes or infixes, and hence cannot be used for Viterbi inference to infer the most likely parse tree for a given infix. Contrastingly our approach can yield parse trees for infixes as well as for prefixes.

6.1   Nederhof and Satta's Algorithm

An *infix* $\mathbf{v}$ in a PCFG G is a substring of a sentence written as $\mathbf{uvw}$ for some terminal sequences $\mathbf{u}$ and $\mathbf{w}$. The infix probability $P_{\text{in}}^{N^1}(\mathbf{v})$ is defined as

$$P_{\text{in}}^{N^1}(\mathbf{v}) = \sum_{\mathbf{u},\mathbf{w}} P_{\mathsf{G}}(\mathbf{uvw})$$

---

[12] The entropy is defined as $-\sum_\tau P(\tau) \log P(\tau)$ where τ is a possible parse tree [2]. In our setting, a common grammar, the universal session grammar, is used for all data sets. So the entropy only depends on the parameters of a PCFG learned from the data set.

```
values(s,[[s,s],[a],[b]]).
:- set_sw(s,[0.4,0.3,0.3]).

infix_pcfg(L):-                         % L : input infix
   build_FA(L),                         % FA asserted in the memory
   assert_last_state(L,End),            % last_state(end(End)) asserted
   start_symbol(C),
   infix_pcfg(0,End,[C]).               % FA transits from state 0 to End
infix_pcfg(S0,S2,[A|R]):-
   ( get_values(A,_) ->                 % A : nonterminal
       msw(A,RHS),                      % use A -> RHS to expand A
       infix_pcfg(S0,S1,RHS)
   ; tr(S0,A,S1) ),                     % state transition by A from S0 to S1
   ( last_state(end(S1)) -> S2=S1       % pseudo success
   ; infix_pcfg(S1,S2,R) ).
infix_pcfg(S,S,[]).
```

Fig. 6. Infix parser DB_2

where u and w range over strings such that uvw is a sentence. According to Nederhof and Satta [10], $P_{\text{in}}^{N^1}(v)$ is computed by first constructing an intersection PCFG $G' = G \cap FA$ of G and a finite automaton FA which accepts every string containing v, and second by computing the sum of probabilities of all sentences derived from G'. The second computation is reduced to solving a set of multi-variate polynomial equations (details omitted).

The problem here is that while their algorithm is completely general, building the intersection PCFG G' contains redundancy. Let $A \to BC$ be a CFG rule in G and $\{s_0, \ldots, s_n\}$ a set of states in FA. To create G', rules of the form $\langle s_i A s_k \rangle \to \langle s_i B s_j \rangle \langle s_j C s_k \rangle$ are constructed for every possible combination of states s_i, s_j, s_k $(0 \leq i, j, k \leq n)$[13] but many of these rules are not used to derive a sentence and need to be removed as useless rules.

6.2   Infix Parsing and Cyclic Explanation Graphs

To avoid building redundant rules by blindly combining states and removing them later, we here propose to introduce parsing to the Nederhof and Satta's algorithm. More concretely, we parse an infix L by the PRISM program in Fig. 6. It is a modification of the prefix parser in Fig. 1 that faithfully simulates the parsing action of the intersection PCFG G'.

---

[13] This is to simulates a state transition of FA made by a string derived from the nonterminal A using $A \to BC$.

This program differs from the prefix parser in that an input infix $\mathbf{w} = w_1 \cdots w_n$ is asserted in the memory as a sequence of state transitions: $\mathtt{tr}(0, w_1, 1), \ldots, \mathtt{tr}(n-1, w_n, n)$, together with other transitions constituting the finite automaton FA. In the program, $\mathtt{tr(S0,A,S1)}$ represents a state transition from S0 to S1 by a word A in the infix. $\mathtt{infix\_pcfg(S0,S2}, \alpha)$ reads that α, a sequence of terminals and nonterminals, spans a terminal sequence which causes a state transition of FA from S0 to S2. Parsing an infix by the infix parser in Fig. 6 yields an explanation graph which is (mostly) cyclic and converted to a set of probability equations just like the case of prefix probability computation. Unlike prefix probability, though, probability equations for an infix are (usually) non-linear and we solve them by Broyden's method, a quasi-Newton method. In this way, we can compute infix probability by way of cyclic explanation graphs. In addition, this program produces the most likely infix parse trees by the Viterbi algorithm on cyclic explanation graphs as explained in Sect. 3.

We experimentally applied the infix method to web session log data and obtained similar results to Fig. 4 (details omitted). However a set of non-linear equations for infix probability computation has multiple roots and a solution given by Broyden's method depends critically on the initial value. Moreover, since Broyden's method is a general solver for non-linear equations, its solution is not necessarily constrained to be between 0 and 1 and actually is often invalid. How to obtain a valid and stable solution of non-linear equations for infix probability computation remains as an open problem.

7 Conclusion

We have proposed new goal and plan recognition methods which are based on prefix and infix probability computation via parse trees in PCFGs. They can identify users' goals and plans behind their incomplete action sequences. A comparative experiment of identifying visitors' goals at a website using three real data sets is conducted using the prefix and infix methods introduced in this paper, the PCFG method that always treats action sequences as complete sentences, the HMM method that uses a mixture of HMMs instead of a mixture of PCFGs, and logistic regression. The result empirically demonstrates the superiority of our approach for long (but incomplete) action sequences.

Another contribution is that our approach removes computational redundancy in Nederhof and Satta's method [10] and also gives infix and prefix parse trees as a side effect. We implemented our approach on top of PRISM2.2[14] which supports prefix and infix parsing and the subsequent probability computation, in particular by automatically solving a set of linear and non-linear probability equations for prefix and infix probability computation respectively.

---

[14] http://rjida.meijo-u.ac.jp/prism/.

References

1. Arlitt, M.F., Williamson, C.L.: Web server workload characterization: the search for invariants. ACM SIGMETRICS Perform. Eval. Rev. **24**, 126–137 (1996)
2. Chi, Z.: Statistical properties of probabilistic context-free grammars. Comput. Linguist. **25**(1), 131–160 (1999)
3. De Raedt, L., Kimmig, A., Toivonen, H.: Problog: a probabilistic prolog and its application in link discovery. IJCAI **7**, 2462–2467 (2007)
4. Goodman, B.A., Litman, D.J.: On the interaction between plan recognition and intelligent interfaces. User Model. User-Adapt. Interact. **2**(1–2), 83–115 (1992)
5. Horvitz, E., Breese, J., Heckerman, D., Hovel, D., Rommelse, K.: The lumiere project: Bayesian user modeling for inferring the goals and needs of software users. In: Proceedings of the Fourteenth Conference on Uncertainty in Artificial Intelligence, pp. 256–265. Morgan Kaufmann Publishers Inc. (1998)
6. Huang, L.: Advanced dynamic programming in semiring and hypergraph frameworks. In: COLING (2008)
7. The Internet Traffic Archive (2001). http://ita.ee.lbl.gov/
8. Jelinek, F., Lafferty, J.D.: Computation of the probability of initial substring generation by stochastic context-free grammars. Comput. Linguist. **17**(3), 315–323 (1991)
9. Lesh, N., Rich, C., Sidner, C.L.: Using plan recognition in human-computer collaboration. In: Kay, J. (ed.) UM99 User Modeling. CISM International Centre for Mechanical Sciences - Courses and Lectures, pp. 23–32. Springer, Vienna (1999)
10. Nederhof, M.J., Satta, G.: Computation of infix probabilities for probabilistic context-free grammars. In: Proceedings of the Conference on Empirical Methods in Natural Language Processing, pp. 1213–1221. Association for Computational Linguistics (2011)
11. Sato, T.: A statistical learning method for logic programs with distribution semantics. In: Proceedings of Internationall Conference on Logic Programming, ILCP 1995 (1995)
12. Sato, T., Kameya, Y.: Parameter learning of logic programs for symbolic-statistical modeling. J. Artif. Intell. Res. **15**, 391–454 (2001)
13. Sato, T., Kameya, Y.: New advances in logic-based probabilistic modeling by PRISM. In: De Raedt, L., Frasconi, P., Kersting, K., Muggleton, S.H. (eds.) Probabilistic Inductive Logic Programming. LNCS (LNAI), vol. 4911, pp. 118–155. Springer, Heidelberg (2008)
14. Sato, T., Meyer, P.: Infinite probability computation by cyclic explanation graphs. Theor. Pract. Logic Program. **15**, 1–29 (2013)
15. Stolcke, A.: An efficient probabilistic context-free parsing algorithm that computes prefix probabilities. Comput. Linguist. **21**(2), 165–201 (1995)
16. Wetherell, C.S.: Probabilistic languages: a review and some open questions. ACM Comput. Surv. (CSUR) **12**(4), 361–379 (1980)

Effectively Creating Weakly Labeled Training Examples via Approximate Domain Knowledge

Sriraam Natarajan[1]([⊠]), Jose Picado[2], Tushar Khot[3], Kristian Kersting[5], Christopher Re[4], and Jude Shavlik[3]

[1] Indiana University, Bloomington, USA
natarasr@indiana.edu
[2] Oregon State University, Corvallis, USA
[3] University of Wisconsin-Madison, Madison, USA
[4] Stanford University, Stanford, USA
[5] Technical University of Dortmund, Dortmund, Germany

Abstract. One of the challenges to information extraction is the requirement of human annotated examples, commonly called gold-standard examples. Many successful approaches alleviate this problem by employing some form of distant supervision, i.e., look into knowledge bases such as Freebase as a source of supervision to create more examples. While this is perfectly reasonable, most distant supervision methods rely on a hand-coded background knowledge that explicitly looks for patterns in text. For example, they assume all sentences containing Person X and Person Y are positive examples of the relation *married(X, Y)*. In this work, we take a different approach – we infer weakly supervised examples for relations from models learned by using knowledge outside the natural language task. We argue that this method creates more robust examples that are particularly useful when learning the entire information-extraction model (the structure and parameters). We demonstrate on three domains that this form of weak supervision yields superior results when learning structure compared to using distant supervision labels or a smaller set of gold-standard labels.

1 Introduction

Supervised learning is one of the popular approaches to information extraction from natural language (NL) text where the goal is to learn relationships between attributes of interest – learn the individuals employed by a particular organization, identifying the winners and losers in a game, etc. There have been two popular forms of supervised learning used for information extraction. First is the classical machine learning approach. For instance, the NIST Automatic Content Extraction (ACE) RDC 2003 and 2004 corpora, has over 1000 documents that have human-labeled relations leading to over 16,000 relations in the documents [12]. ACE systems use textual features – lexical, syntactic and semantic – to learn mentions of target relations [21,27]. But pure supervised approaches are quite limited in scalability due to the requirement of high

© Springer International Publishing Switzerland 2015
J. Davis and J. Ramon (Eds.): ILP 2014, LNAI 9046, pp. 92–107, 2015.
DOI: 10.1007/978-3-319-23708-4_7

quality labels, which can be very expensive to obtain for most NL tasks. An attractive second approach is *distant supervision*, where labels of relations in the text are created by applying a heuristic to a common knowledge base such as Freebase [12,19,23]. The quality of these labels are crucially dependent on the heuristic used to map the relations to the knowledge base. Consequently, there have been several approaches that aim to improve the quality of these labels ranging from multi-instance learning [5,19,22] to using patterns that frequently appear in the text [23]. As noted by Riedel et al. [19], the distant supervision assumption can be too strong, particularly when the source used for labeling the examples is external to the learning task at hand.

Hence, we use the probabilistic logic formalism called *Markov Logic Networks* [4] to perform *weak supervision* [1] to create more examples. Instead of directly obtaining the labels from a different source, we perform inference on *outside knowledge* (i.e., knowledge not explicitly stated in the corpus) to create sets of entities that are "potential" relations. This outside knowledge forms the *context MLN* – CMLN – to reflect that they are non-linguistic models. An example of such knowledge could be that "home team are more likely to win a game". Note that this approach enables the domain expert to write rules in first-order logic so that the knowledge is not specific to any particular textural wording but is general knowledge about the world (in our example, about the games played). During the information extraction (IE) phase, unlabeled text are then parsed through some entity resolution parser to identify potential entities. These entities are then used as queries to the CMLN which uses data from non-NLP sources to infer the posterior probability of relations between these entities. These inferred relations become the probabilistic examples for IE. This is in contrast to distant supervision where statistical learning is employed at "system build time" to construct a function from training examples.

Our hypothesis – which we verify empirically – is that the use of world knowledge will help in learning from NL text. This is particularly true when there is a need to learn a model without any prior structure (e.g. a MLN) since the number of examples needed can be large. These weakly supervised examples can augment the gold-standard examples to improve the quality of the learned models. So far, the major hurdle to learning structure in IE is the large number of features leading to increased complexity in the search [17]. Most methods use a prior designed graphical model and **only** *learn the parameters*. A key issue with most structure-learning methods is that, when scoring every candidate structure, parameter learning has to be performed in the inner loop. We, on the other hand, employ an algorithm based on *Relational Functional Gradient Boosting* (RFGB) [8,9,13] for learning **the structure**. It must be emphasized clearly that the main contribution of the paper is not the learning algorithm, but instead is presenting a method for generation of weakly supervised examples to augment the gold standard examples.

We then employ RFGB in three different tasks:

1. Learning to jointly predict game winners and losers from NFL news articles[1]. We learned from 50 labeled documents and used 400 unlabeled documents. For the unlabeled documents, we used a common publicly available knowledge base such as Freebase to perform inference on the game winners and losers
2. Classifying documents as either *football* or *soccer* articles (using the American English senses of these words) from a data set that we created
3. Comparing our weak supervision approach to the distant supervision approach on a standard NYT data set[2].

We perform 5-fold cross validation on these tasks and show that the proposed approach outperforms learning only from gold-standard data.

When we can bias the learner with examples created from commonsense knowledge, we can distantly learn model structure. Because we have more training examples than the limited supply of gold-standard examples and they are of a higher quality than traditional distant labeling, the proposed approach allows for a better model to be learned. Our algorithm has two phases:

1 *Weak supervision phase*, where the goal is to use commonsense knowledge (CMLN). This CMLN could contain clauses such as "Higher ranked teams are more likely to win".
2 *Information extraction phase*, where the noisy examples are combined with some "gold-standard" examples and a relational model is learned using RFGB on textual features.

The potential of such advice giving method is not restricted to NL tasks and is more broadly applicable. For instance, this type advice can be used for labeling tasks [24] or to shape rewards in reinforcement learning [2] or to improve the number of examples in a medical task. Such advice can also be used to provide guidance to a learner in unforseen situations [11].

2   Related Work

Distant Supervision: Our approach is quite similar to the *distant supervision* [1,12] that generate training examples based on external knowledge bases. Sentences in which any of the related entities are mentioned, are considered to be positive training examples. These examples along with the few annotated examples are provided to the learning algorithm. These approaches assume that the sentences that mention the related entities probably express the given relation. Riedel et al. [19] relax this assumption by introducing a latent variable for each mention pair to indicate whether the relation is mentioned or not. This work was further extended to allow overlapping relations between the same pair of entities (e.g. Founded(Jobs, Apple) and CEO-of(Jobs, Apple)) by using a multi-instance multi-label approach [5,22]. We employ a model based on non-linguistic knowledge to generate the distant supervision examples. Although we

---

[1] LDC catalog number LDC2009E112.
[2] LDC catalog number LDC2008T19.

rely on a knowledge base to obtain the relevant input relations for our CMLN model, one can imagine tasks where such relations are available as inputs or extracted earlier in the pipeline.

Statistical Relational Learning: Most NLP approaches define a set of features by converting structured output such as parse trees, dependency graphs, etc. to a flat feature vector and use propositional methods such as logistic regression. Recently, there has been a focus of employing Statistical Relational models that combine the expressiveness of first-order logic and the ability of probability theory to model uncertainty. Many tasks such as BioNLP [10] and TempEval [25] have been addressed [16,18,26] using SRL models, namely Markov Logic Networks (MLNs) [4]. But these approaches still relied on generating features from structured data. Sorower et al. [20] use a similar concept in spirit where they introduce a mention mode that models the probability of facts mentioned in the text and use a EM algorithm to learn MLN rules to achieve this. We represent the structured data (e.g. parse trees) using first-order logic and use the RFGB algorithm to learn the structure of Relational Dependency Networks (RDN) [14]. Relational Dependency Networks (RDNs) are SRL models that consider a joint distribution as a product of conditional distributions. One of the important advantages of RDNs is that the models are allowed to be cyclic. As shown in the next section, we use MLNs to specify the weakly supervised world knowledge.

3   Structure Learning for Information Extraction Using Weak Supervision

One of the most important challenges facing many natural language tasks is the paucity of "gold standard" examples. Our proposed method, shown in Fig. 1, has two distinct phases: *weak supervision phase* where we create weakly supervised examples based on commonsense knowledge and *information extraction phase* where we learn the structure and parameters of the models that predict relations using textual features.

3.1   Weak Supervision Phase

We now explain how our first phase addresses the key challenge of obtaining additional training examples. As mentioned earlier, the key challenge is obtaining annotated examples. To address this problem, we employ a method that is commonly taken by humans. For instance, consider reading a newspaper sports section about a particular sport (say the NFL). We have an inherent *inductive bias* – we expect a high ranked team (particularly if it plays at home) to win. In other words, we rarely expect "upsets". We aim to formalize this notion by employing a model that captures this inductive bias to label in addition to gold standard examples.

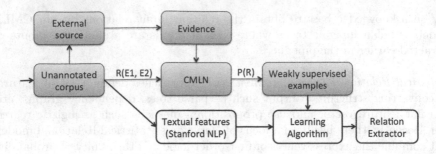

Fig. 1. Flowchart of our method. The top-half represents the weak supervision phase where we generate the examples using the CMLN and facts from an external source. The bottom-half represents the information extraction phase where we learn a SRL model using the weakly supervised and gold standard examples.

We employ MLNs to capture this world knowledge. MLNs [4] are relational undirected models where first-order logic formula correspond to the cliques of a Markov network and formula weights correspond to the clique potentials. A MLN can be instantiated as a Markov network with a node for each ground predicate (atom) and a clique for each ground formula. All groundings of the same formula are assigned the same weight. So the joint probability distribution over all atoms is

$$P(X = x) = \frac{1}{Z} \exp \left(\sum_i w_i n_i(x) \right) \tag{1}$$

where $n_i(x)$ is the number of times the ith formula is satisfied by possible world x and Z is a normalization constant. Intuitively, a possible world where formula f_i is true one more time than a different possible world is e^{w_i} times as probable, all other things being equal. There have been several weight learning, structure learning and inference algorithms proposed for MLNs. MLNs provide an easy way for domain's expert to specify the background knowledge and effective algorithms exist for learning the weights of these clauses and perform inference. We use the scalable Tuffy system [15] to perform inference. One of the key attractions of Tuffy is that it can scale to millions of documents.

Our proposed approach for weak supervision is presented in Fig. 2. The first step is to design a MLN that captures domain knowledge, called as *CMLN*. For the NFL domain, some rules that we used are shown in Table 1. For example, "Home team is more likely to win the game" (first two clauses), "High ranked team is more likely to win the game" (last two rules). Note that the rules are written without having the knowledge base in mind. These rules are simply written by the domain's expert and they are softened using a knowledge base such as Wikipedia. The resulting weights are presented in the left column of the table. We used the games played in the last 20 years to compute these weights.

While it is possible to learn these weights from data (for instance using previously played NFL games), we set the weights based on the log-odds for CMLN. For instance, for NFL domain, we set the weights of the clauses by

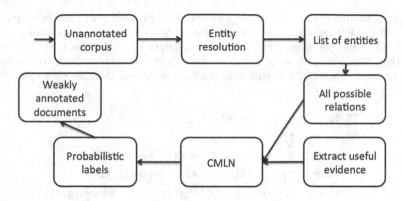

Fig. 2. Steps involved in creation of weakly supervised examples.

considering the previously played NFL games[3]. We looked at the number of games played, the total number of times a home team won and the total number of times a higher ranked[4] team won, etc. If a home team won 10 times roughly more compared to away team, the weights were set to be $log(10) = 1$ for the rule about home team winning more often. Using this data, we set the weights of the MLN clauses as shown in Table 1. This is an approach that has been taken earlier when employing the use of MLNs [6, 7]. In our experiments, we found that the results are not very sensitive to the exact MLN weights as long as the order of the rule weights is preserved.

Note that one could simply define a higher ranking using the following MLN clause where t denotes a team, r its rank, y the year of the ranking and hR the higher rank: $\infty \quad rank(t1, r1, y), rank(t2, r2, y), t1! = t2, r1 < r2 \rightarrow hR(t1, t2, y)$. We argue that this is one of the major features of using common-sense knowledge. The relative merits between the different rules can be judged reasonably even though a domain's expert may not fully understand the exact impact of the weights. It is quite natural in several tasks, as we show empirically, to set "reasonable" weights.

The next step is to create weakly supervised learning examples. We identify interesting (unannotated) documents – for example, sport articles from different news web sites. We use *Stanford NLP* toolkit to perform entity resolution to identify the potential teams, games and year in the document. Then, we use the CMLN to obtain the posterior probability on the relations being true between entities mentioned in the same sentence – for example, game winner and loser relations. Note that to perform inference, evidence is required. Hence, we use the games that have been played between the two teams (again from previously played games that year) to identify the home, away and ranking of the teams. We used the rankings at the start of the year of the game as a pseudo reflection of the relative rankings between the teams.

---

[3] We obtained from Pro-Football-Reference http://www.pro-football-reference.com/.
[4] According to http://www.nfl.com/.

Table 1. A sample of CMLN clauses used for the NFL task with their corresponding weights on the left column. t denotes a team, g denotes a game, y denotes the year, $tInG$ denotes that the team t plays in game g, $hR(t1, t2, y)$ denotes that $t1$ is ranked higher than $t2$ in year y. A weight of ∞ means that the clause is a "hard" constraint on the set of possible worlds, a weight different from ∞ means that the clause is a "soft" contraint.

0.33	home(g, t) $\rightarrow$ winner(g, t)
0.33	away(g, t) $\rightarrow$ loser(g, t)
∞	exist t2 winner(g, t1), t1 != t2 $\rightarrow$ loser(g, t2)
∞	exist t2 loser(g, t1), t1 != t2 $\rightarrow$ winner(g, t2)
0.27	tInG(g, t1), tInG(g, t2), hR(t1, t2, y) $\rightarrow$ winner(g, t1)
0.27	tInG(g, t1), tInG(g, t2), hR(t1, t2, y) $\rightarrow$ loser(g, t2)

Recall that the results of inference are the posterior probabilities of the relations being true between the entities extracted from the same sentence and they are used for annotations. One simple annotation scheme is using the MAP estimate (i.e., if the probability of a team being a winner is greater than the probability of being the loser, the relation becomes a positive example for winner and a negative example for loser). An alternative would be to use a method that directly learns from probabilistic labels. Choosing the MAP would make a strong commitment about several examples on the borderline. Since our world knowledge is independent of the text, it may be the case that for some examples perfect labeling is not easy. In such cases, using a softer labeling method might be more beneficial. Now these weakly supervised examples are ready for our next step – *information extraction*.

3.2 Learning for Information Extraction

Once the weakly supervised examples are created, the next step is inducing the relations. We employ the procedure from Fig. 3. We run both the gold standard and weakly supervised annotated documents through Stanford NLP toolkit to create linguistic features. Once these features are created, we run the RFGB algorithm [13]. This allows us to create a joint model between the target relations, for example, game winner and losers. We now briefly describe the adaptation of RFGB to this task.

Let the training examples be of the form $(\mathbf{x}_i, y_i)$ for $i = 1, ..., N$ and $y_i \in \{1, ..., K\}$. $\mathbf{x}$ denotes the features which in our case are lexical and syntactic features and ys correspond to target (game winners and loser) relations. Relational models tend to consider training instances as "mega examples" where each example represents all instances of a particular group. We consider each document to be a mega example i.e., we do not consider cross document learning.

The goal is to fit a model $P(y|\mathbf{x}) \propto e^{\psi(y,\mathbf{x})}$ for every target relation y. Functional gradient ascent starts with an initial potential ψ_0 and iteratively adds

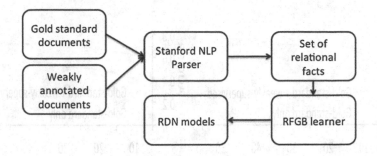

Fig. 3. Steps involved in learning using probabilistic examples.

gradients Δ_i. Δ_m is the functional gradient at episode m and is

$$\Delta_m = \eta_m \times E_{x,y}[\partial/\partial\psi_{m-1} log\, P(y|x; \psi_{m-1})] \tag{2}$$

where η_m is the learning rate. Dietterich *et al.* [3] suggested evaluating the gradient for every training example and fitting a regression tree to these derived examples $([(x_i, y_i), \Delta_m(y_i; x_i)])$.

In our formalism, y corresponds to target relations, for example *gameWinner* and *gameLoser* relations between a team and game mentioned in a sentence. x corresponds to all the relational facts produced by Stanford NLP toolkit – lexical features, such as base forms of words, part-of-speech tags, word lemmas and entity types, and syntactic features such as phrase chunks, phrase types, parse trees and dependency paths. To learn the model for a relation, say *gameWinner*, we start with an initial model ψ_0 (uniform distribution). Next, we calculate the gradients for each example as the difference between the true label and current predicted probability. Then, we learn a relational regression tree to fit the regression examples and add it to the current model. We now compute the gradients based on the updated model and repeat the process. In every subsequent iteration, we *fix* the errors made by the model. For further details, we refer to Natarajan et al. [13].

Since we use a probabilistic model to generate the weakly supervised examples, our training examples will have probabilities associated with them based on the predictions from CMLN. We extend RFGB to handle probabilistic examples by defining the loss function as the KL-divergence[5] between the observed probabilities (shown using P_{obs}) and predicted probabilities (shown using P_{pred}). The gradients for this loss function is the difference between the observed and predicted probabilities.

$$\Delta_m(x) = \frac{\partial}{\partial\psi_{m-1}} \sum_{\hat{y}} P_{obs}(y = \hat{y}) \log\left(\frac{P_{obs}(y = \hat{y})}{P_{pred}(y = \hat{y}|\psi_{m-1})}\right)$$

$$= P_{obs}(y = 1) - P_{pred}(y = 1|\psi_{m-1})$$

---

[5] $D_{KL}(P;Q) = \sum_y P(y) log(P(y)/Q(y))$.

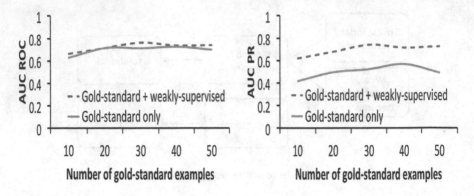

Fig. 4. Results of predicting winners and losers in NFL : (a) AUC ROC. (b) AUC PR.

Hence the key idea in our work is to use probabilistic examples that we obtain from the weakly supervised phase as input to our structure learning phase along with gold standard examples and their associated documents. Then a RDN is induced by learning to predict the target relations jointly, using features created by the Stanford NLP toolkit. Since we are learning a RDN, we do not have to explicitly check for acyclicity. We chose to employ RDNs as they have been demonstrated to have the state-of-the-art performance in many relational and noisy domains [13]. We use modified ordered Gibbs sampler [14] for inference.

4   Experimental Results

We now present the results of empirically validating our approach on three domains. In the first two domains, we compare the use of augmenting with weakly supervised examples against simply using the gold standard examples. In the third domain, we compare the examples generated using our weak supervision approach against distant supervision examples on a standard data set. We also compare against one more distant supervision method that is the state-of-the-art on this data set.

4.1   NFL Relation Extraction

The first data set on which we evaluate our method is the National Football League (NFL) data set from LDC[6] that consists of articles of NFL games from past two decades. The goal is to identify relations such as *winner* and *loser*. For example, consider the text, "Packers defeated Cowboys 28 − 14 in Saturday's Superbowl game". The goal is to identify *Greenbay* and *Dallas* as the winner and loser respectively. The corpus consists of articles, some of which are annotated with target relations. We consider only articles that have annotations of positive

---

[6] http://www.ldc.upenn.edu.

examples. There were 66 annotations of the relations. We used 16 of these annotations as the test set and performed training on the rest. In addition to the gold
standard examples, we used articles from the NFL website[7] for weak supervision. We used the MLN presented in Table 1 for inferring the weakly supervised
examples.

The goal is to evaluate the impact of the weakly supervised examples. We
used 400 weakly supervised examples as the results did not improve beyond using
400 examples. We varied the number of gold standard examples while keeping
the number of weakly supervised examples constant. We compared against using
only gold standard examples. The results were averaged over 5 runs of random
selection of gold standard examples. We measured the area under curves for
both ROC and PR curves. Simply measuring the accuracy on the test set will
not suffice as predicting the majority class can lead in high performance. Hence
we present AUC. The results are presented in Fig. 4 where the performance
measure is presented by varying the number of gold standard examples. As
can be seen, in both metrics, the weakly supervised examples improve upon
the usage of gold standard examples. The use of weakly supervised examples
allows a more accurate performance with a small number of examples, a steeper
learning curve and in the case of PR, convergence to a higher value. It should
be mentioned that for every point in the graph, we sample the gold standard
examples from a fixed set of examples and the only difference is whether there are
any weakly supervised examples added. For example, when plotting the results
of 10 examples, the set of gold standard examples is the same for every run. For
the blue dashed curve, we add 400 more weakly supervised examples and this is
repeated for 5 runs in which the 10 gold examples are drawn randomly.

We also performed t-tests on all the points of the PR and ROC curves. For
the PR curves, the use of weakly supervised learning yields statistically superior
performance over the gold standard examples for all the points on the curves
(with p-value < 0.05). For the ROC curves, significance occurs when using 10
and 30 examples. Since PR curves are more conservative than ROC curves, it is
clear that the use of these weakly supervised examples improves the performance
of the structure learner significantly. To understand whether weak supervision
helps, we randomly assigned labels to the 400 examples (instead of using weak
supervision). When combined with 50 gold examples, the performance decreased
dramatically with AUC values being less than 0.6. This clearly shows that the
weakly supervised labels help when learning the structure. We also used the
MAP estimates from the MLN to label the weakly supervised examples. The
performance was comparable to that of using random labels labels across all
runs with an average AUC-ROC of 0.6.

4.2   Document Classification

As a second experiment, we created another data set where the goal is to classify documents either as being *football(American)* or *soccer* articles. The target

---

[7] http://www.nfl.com.

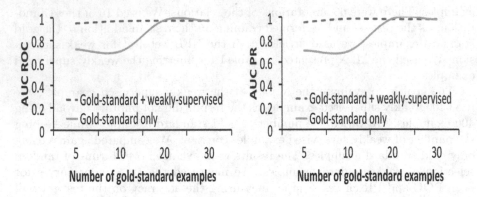

Fig. 5. Results of classifying football and soccer articles in the document classification domain: **(Left)** AUC ROC. **(Right)** AUC PR.

relation is the type of the article. We extracted 30 football articles from the NFL website[8] and 30 soccer articles from the English Premier League (EPL) website[9] and annotated them manually as being football and soccer respectively. We used only the first paragraph of the articles for learning the models since it appeared that enough information is present in the first paragraph for learning an useful model. We purposefully annotated a small number of articles as the goal is to analyze performance under a small number of gold standard examples. In addition, we used 45 articles for weak supervision (as with the previous case, the performance did not improve with more weakly supervised articles). We used rules such as, "If a soccer league and a soccer team are mentioned, then it is a soccer game", "If a football league and a football team are mentioned, then it is a football game", "EPL teams play soccer", "NFL teams play football", "If the scores of both teams are equal or greater than 10, then it is a football game", "If the scores of both teams are less than 10, then it is a soccer game", "If the scores of both teams are 0, then it is a soccer game". These are presented in Table 2.

All the rules mentioned are essentially considered as "soft" rules. The weights of these rules were simply set to 100, 10, 1 to reflect the log-odds. During the weak supervision phase, we used the entities mentioned in the documents as queries to CMLN to predict the type. These predictions (probabilities) become the weak supervision for the learning phase. As with NFL, we measured the AUC ROC and PR values by varying the number of gold standard examples. Again, in each run, to maintain consistency, we held the gold standard examples to be constant and simply added the weakly supervised examples. The results are presented in Fig. 5. The resulting figures show that as with the earlier case, weak supervision helps improve the performance of the learning algorithm. We get a jump start and a steeper learning curve in this case as well. Again, the results are statistically

---

[8] http://www.nfl.com.
[9] http://www.premierleague.com.

Table 2. CMLN clauses used for document classification. a denotes an article, t denotes a team, l denotes a league, te denotes that t is a team mentioned in article a, le denotes that l is a league mentioned in article a, $tInL$ denotes that the team t plays in league l, sL denotes that l is a soccer league, fL denotes that l is a football league, sco denotes that t had a final score of s in a.

100	te(a,t), le(a,l), tInL(t,l), sL(l) $\rightarrow$ soccer(a)
100	te(a,t), le(a,l), tInL(t,l), fL(l) $\rightarrow$ football(a)
10	te(a,t), sL(l), tInL(t,l) $\rightarrow$ soccer(a)
10	te(a,t), fL(l), tInL(t,l) $\rightarrow$ football(a)
1	te(a,t1), te(a,t2), sco(a,t1,s1), sco(a,t2,s2), s1 $>=$ 10, s2 $>=$ 10 $\rightarrow$ football(a)
1	te(a,t1), te(a,t2), sco(a,t1,s1), sco(a,t2,s2), s1 $<$ 10, s2 $<$ 10 $\rightarrow$ soccer(a)
1	te(a,t1), te(a,t2), sco(a,t1,s1), sco(a,t2,s2), s1 $=$ 0, s2 $=$ 0 $\rightarrow$ soccer(a)

significant for small number of gold standard examples. The experiment proves that adding probabilistic examples as weak supervision improves performance. Note that with more gold standard examples, the performance decreases due to overfitting. But with weak supervision, since the examples are probabilistic, this issue is avoided – another observation that the use of weaker examples aids the learner.

4.3   NY Times Relation Extraction

The question that we explicitly aim to ask in this experiment is: How does this method of weak supervision compare against a distant supervision method to create the examples for structure-learning? We used the corpus created by Riedel et al. [19] by aligning the NYT corpus between the years 2005 and 2007 with Freebase[10] relations. The same corpus was subsequently used by Hoffmann et al. [5] and Surdeanu et al. [22]. We restricted our attention to only the *nationality* relation. We performed 5 runs with 70 examples for training and 50 examples for testing. Surdeanu et al. [22] exhibit the state-of-the-art performance in this task. But they **only learn the parameters** of the graphical model while we learn the entire model. Hence we did not compare against their method.

To generate probabilistic examples, we designed an MLN and set its weights to predict the nationality of a person based on his or her place of birth and location of residence. The MLN is presented in Table 3. We queried Freebase for the place of birth of a person and the places the person has lived in, and used this information as evidence. We used rules such as "If a person was born in a country, then the person's nationality is that country" and "If a person has lived in a country, then the person's nationality is that country" (first and third clauses). Because the place of birth or location of residence provided by Freebase are not always countries, we also queried Freebase for the countries in which the locations are contained. We also used rules rules such as "If a person was born

---

[10] http://www.freebase.com/.

Table 3. CMLN used for the NYT relation extraction. p denotes a person, l denotes a location, c denotes a country.

1	place_of_birth(p, c) $\rightarrow$ nationality(p, c)
1	place_of_birth(p, l), contained_by(l, c) $\rightarrow$ nationality(p, c)
0.1	place_lived(p, c) $\rightarrow$ nationality(p, c)
0.1	place_lived(p, l), contained_by(l, c)$\rightarrow$ nationality(p, c)

Table 4. Weak (MLN) & distant supervision (Freebase) results.

	AUC ROC	AUC PR
MLN	**0.53**	**0.56**
Freebase	0.48	0.52

in a location (e.g., a city) and that location is contained in a country, then the person's nationality is that country" and "If a person has lived in a location and that location is contained in a country, then the person's nationality is that country" (second and fourth clauses).

We queried the MLN to obtain the posterior probabilities on the *nationality* relation between each person and country. Finally we used the person and country entities, as well as the corresponding posterior probabilities, to create new probabilistic examples for RFGB.

We compare our examples to the distant supervision examples obtained from Freebase. To obtain the distant supervision examples, we queried Freebase for the nationality of a person and looked for sentences in the dataset that contained each pair of entities. We used the same structure-learning approach for the weak supervision (obtained from the MLN) and distant supervision (obtained from Freebase) examples. The results comparing both settings are presented in Table 4. As can be seen, the weak supervision (MLN) gets better performance than the distant supervision (Freebase) in both AUC ROC and AUC PR, which means that the supervision provided by the CMLN results in examples of higher quality. This observation is similar to the one made by Riedel et al. [19], as the Freebase is not necessarily the source of the NYT corpus (or vice-versa), the use of this knowledge base (distant supervision) makes a stronger assumption than our inferred "soft" examples. Using soft (probabilistic) labels is beneficial as opposed to a fixed label of a relation as the latter can possibly overfit.

We also compared the best F1 value of our approach against the state-of-the-art distant supervision approach [22], which only learns the parameters for a hand-written structure. This is to say that the approach of Surdeanu et al. [22] assumes that the structure of the model is provided and hence simply learns the parameters of the model. On the other hand, our learning algorithm learns the model as well as its parameters. We obtained the code[11] for Surdeanu et al.'s

---

[11] http://nlp.stanford.edu/software/mimlre.shtml.

Table 5. Best F1 value for MIML-RE and our approach for the *nationality* relation.

	Best F1
MIML-RE	0.16
MLN	0.14

approach and modified it to focus only on the *nationality* relation in both learning and inference. The best F1 values for this approach (MIML-RE) and our approach (MLN) are presented in Table 5. We report the best F1 values as this metric was reported in their earlier work.

5   Conclusion

One of the key challenges for applying learning methods in many real-world problems is the paucity of good quality labeled examples. While semi-supervised learning methods have been developed, we explore another alternative method of weak supervision – where the goal is to create examples of a quality that can be relied upon. We considered the NLP tasks of relation extraction and document extraction to demonstrate the usefulness of the weak supervision. Our key insight is that weak supervision can be provided by a "domain" expert instead of a "NLP" expert and thus the knowledge is independent of the underlying problem but is close to the average human thought process – for example, sports fans. We are exploiting domain knowledge that the authors of articles assume their readers already know and hence the authors do not state it. We used the weighted logic representation of Markov Logic networks to model the expert knowledge, infer the relations in the unannotated articles, and adapted functional gradient boosting for predicting the target relations. Our results demonstrate that our method significantly improves the performance thus reducing the need for gold standard examples.

Our proposed method is closely related to distant supervision methods. So it will be an interesting future direction to combine the distant and weak supervision examples for structure learning. Combining weak supervision with advice taking methods [2,11,24] is another interesting direction. This method can be seen as giving advice about the examples, but AI has a long history of using advice on the model, the search space and examples. Hence, combining them might lead to a strong knowledge based system where the knowledge can be provided by a domain expert and not a AI/NLP expert. We envision that we should be able to "infer" the world knowledge from knowledge bases such as Cyc or ConceptNet and employ them to generate the weak supervision examples. Finally, it is important to evaluate the proposed model in similar tasks.

Acknowledgements. Sriraam Natarajan, Tushar Khot, Jose Picado, Chris Re, and Jude Shavlik gratefully acknowledge support of the DARPA Machine Reading Program and DEFT Program under the Air Force Research Laboratory (AFRL) prime contract no. FA8750-09-C-0181 and FA8750-13-2-0039 respectively. Any opinions, findings, and conclusion or recommendations expressed in this material are those of the authors and do not necessarily reflect the view of the DARPA, AFRL, or the US government. Kristian Kersting was supported by the Fraunhofer ATTRACT fellowship STREAM and by the European Commission under contract number FP7-248258-First-MM.

References

1. Craven, M., Kumlien, J.: Constructing biological knowledge bases by extracting information from text sources. In: ISMB (1999)
2. Devlin, S., Kudenko, D., Grzes, M.: An empirical study of potential-based reward shaping and advice in complex, multi-agent systems. Adv. Complex Syst. **14**(2), 251–278 (2011)
3. Dietterich, T.G., Ashenfelter, A., Bulatov, Y.: Training conditional random fields via gradient tree boosting. In: ICML (2004)
4. Domingos, P., Lowd, D.: Markov Logic: An Interface Layer for AI. Morgan & Claypool, San Rafael (2009)
5. Hoffmann, R., Zhang, C., Ling, X., Zettlemoyer, L., Weld, D.S.: Knowledge-based weak supervision for information extraction of overlapping relations. In: ACL (2011)
6. Jain, D.: Knowledge engineering with Markov logic networks: a review. In: KR (2011)
7. Kate, R., Mooney, R.: Probabilistic abduction using Markov logic networks. In: PAIR (2009)
8. Kersting, K., Driessens, K.: Non-parametric policy gradients: a unified treatment of propositional and relational domains. In: ICML (2008)
9. Khot, T., Natarajan, S., Kersting, K., Shavlik, J.: Learning Markov logic networks via functional gradient boosting. In: ICDM (2011)
10. Kim, J., Ohta, T., Pyysalo, S., Kano, Y., Tsujii, J.: Overview of BioNLP'09 shared task on event extraction. In: BioNLP Workshop Companion Volume for Shared Task (2009)
11. Kuhlmann, G., Stone, P., Mooney, R.J., Shavlik, J.W.: Guiding a reinforcement learner with natural language advice: initial results in robocup soccer. In: AAAI Workshop on Supervisory Control of Learning and Adaptive Systems (2004)
12. Mintz, M., Bills, S., Snow, R., Jurafsky, D.: Distant supervision for relation extraction without labeled data. In: ACL and AFNLP (2009)
13. Natarajan, S., Khot, T., Kersting, K., Guttmann, B., Shavlik, J.: Gradient-based boosting for statistical relational learning: the relational dependency network case. Mach. Learn. **86**(1), 25–56 (2012)
14. Neville, J., Jensen, D.: Relational dependency networks. In: Getoor, L., Taskar, B. (eds.) Introduction to Statistical Relational Learning, pp. 653–692. MIT Press, Cambridge (2007)
15. Niu, F., Ré, C., Doan, A., Shavlik, J.W.: Tuffy: scaling up statistical inference in Markov logic networks using an RDBMS. PVLDB **4**(6), 373–384 (2011)
16. Poon, H., Vanderwende, L.: Joint inference for knowledge extraction from biomedical literature. In: NAACL (2010)
17. Raghavan, S., Mooney, R.: Online inference-rule learning from natural-language extractions. In: International Workshop on Statistical Relational AI (2013)
18. Riedel, S., Chun, H., Takagi, T., Tsujii, A J.: Markov logic approach to biomolecular event extraction. In: BioNLP (2009)

19. Riedel, S., Yao, L., McCallum, A.: Modeling relations and their mentions without labeled text. In: Balcázar, J.L., Bonchi, F., Gionis, A., Sebag, M. (eds.) ECML PKDD 2010, Part III. LNCS, vol. 6323, pp. 148–163. Springer, Heidelberg (2010)
20. Sorower, S., Dietterich, T., Doppa, J., Orr, W., Tadepalli, P., Fern, X.: Inverting grice's maxims to learn rules from natural language extractions. In: NIPS, pp. 1053–1061 (2011)
21. Surdeanu, M., Ciaramita, M.: Robust information extraction with perceptrons. In: NIST ACE (2007)
22. Surdeanu, M., Tibshirani, J., Nallapati, R., Manning, C.: Multi-instance multi-label learning for relation extraction. In: EMNLP-CoNLL (2012)
23. Takamatsu, S., Sato, I., Nakagawa, H.: Reducing wrong labels in distant supervision for relation extraction. In: ACL (2012)
24. Torrey, L., Shavlik, J., Walker, T., Maclin, R.: Transfer learning via advice taking. In: Koronacki, J., Raś, Z.W., Wierzchoń, S.T., Kacprzyk, J. (eds.) Advances in Machine Learning I. SCI, vol. 262, pp. 147–170. Springer, Heidelberg (2010)
25. Verhagen, M., Gaizauskas, R., Schilder, F., Hepple, M., Katz, G., Pustejovsky, J.: SemEval-2007 task 15: TempEval temporal relation identification. In: SemEval (2007)
26. Yoshikawa, K., Riedel, S., Asahara, M., Matsumoto, Y.: Jointly identifying temporal relations with Markov logic. In: ACL and AFNLP (2009)
27. Zhou, G., Su, J., Zhang, J., Zhang, M.: Exploring various knowledge in relation extraction. In: ACL (2005)

Learning Prime Implicant Conditions from Interpretation Transition

Tony Ribeiro[1]([⊠]) and Katsumi Inoue[1,2]

[1] The Graduate University for Advanced Studies (Sokendai),
2-1-2 Hitotsubashi, Chiyoda-ku, Tokyo 101-8430, Japan
tony_ribeiro@nii.ac.jp
[2] National Institute of Informatics,
2-1-2 Hitotsubashi, Chiyoda-ku, Tokyo 101-8430, Japan
inoue@nii.ac.jp

Abstract. In a previous work we proposed a framework for learning normal logic programs from transitions of interpretations. Given a set of pairs of interpretations (I, J) such that $J = T_P(I)$, where T_P is the immediate consequence operator, we infer the program P. Here we propose a new learning approach that is more efficient in terms of output quality. This new approach relies on specialization in place of generalization. It generates hypotheses by *specialization* from the most general clauses until no negative transition is covered. Contrary to previous approaches, the output of this method does not depend on variables/transitions ordering. The new method guarantees that the learned rules are minimal, that is, the body of each rule constitutes a prime implicant to infer the head.

Keywords: Dynamical systems · Boolean networks · Attractors · Supported models · Learning from interpretation · Inductive logic programming

1 Introduction

In recent years, there has been a notable interest in the field of Inductive Logic Programming (ILP) to learn from system state transitions as part of a wider interest in learning the dynamics of systems [2,4,8,14]. Learning system dynamics has many applications in multi-agent systems, robotics and bioinformatics alike. Knowledge of system dynamics can be used by agents and robots for planning and scheduling. In bioinformatics, learning the dynamics of biological systems can correspond to the identification of the influence of genes and can help to design more efficient drugs. In some previous works, state transition systems are represented with logic programs [6,9], in which the state of the world is represented by an Herbrand interpretation and the dynamics that rule the environment changes are represented by a logic program P. The rules in P specify the next state of the world as an Herbrand interpretation through the *immediate consequence operator* (also called the T_P *operator*) [1,18]. With such

© Springer International Publishing Switzerland 2015
J. Davis and J. Ramon (Eds.): ILP 2014, LNAI 9046, pp. 108–125, 2015.
DOI: 10.1007/978-3-319-23708-4_8

a background, Inoue *et al.* [8] have recently proposed a framework to learn logic programs from traces of interpretation transitions (LFIT). The learning setting of this framework is as follows. We are given a set of pairs of Herbrand interpretations (I, J) as positive examples such that $J = T_P(I)$, and the goal is to induce a *normal logic program* (NLP) P that realizes the given transition relations. In [8], the authors showed one of the possible usages of LFIT: **LF1T**, *learning from 1-step transitions*. In that paper, an algorithm is proposed to iteratively learn an NLP that realizes the dynamics of the system by considering step transitions one by one. The iterative character of **LF1T** has applications in bioinformatics, cellular automata, multi-agent systems and robotics.

In this paper, our main concern is the minimality of the rules and the NLPs learned by LF1T. Our goal is to learn all minimal conditions that imply a variable to be true in the next state, e.g. all prime implicant conditions. In bioinformatics, for a gene regulatory network, it corresponds to all minimal conditions for a gene to be activated/inhibited. It can be easier and faster to perform model checking on Boolean networks represented by a compact NLP than the set of all state transitions. Knowing the minimal conditions required to perform the desired state transitions, a robot can optimize its actions to achieve its goals with less energy consumption. From a technical point of view, for the sake of memory usage and reasoning time, a small NLP could also be preferred in multi-agent and robotics applications. For this purpose, we propose a new version of the **LF1T** algorithm based on specialization. Specialization is usually considered the dual of generalization in ILP [11–13]. Where generalization occurs when a hypothesis does not explain a positive example, specialization is used to refine a hypothesis that implies a negative example.

In [7], prime implicants are defined for DNF formula as follows: a clause C, implicant of a formula ϕ, is prime if and only if none of its proper subset $S \subset C$ is an implicant of ϕ. In this work, explanatory induction is considered, while in our approach prime implicants are defined in the LFIT framework. Knowing the Boolean functions, prime implicants could be computed by Tisons consensus method [17] and its variants [10]. The novelty of our approach, is that we compute prime implicants incrementally during the learning of the Boolean function. In [8,16], **LF1T** uses resolution techniques to generalize rules and reduces the size of the output NLP. This technique generates hypotheses by *generalization* from the most specific clauses until every positive transitions are covered. Compared to previous **LF1T** algorithms, the novelty of our new approach is that it generates hypotheses by *specialization* from the most general clauses until no negative transition is covered. The main weak point of the previous **LF1T** algorithms is that the output NLPs depends on variable/transition ordering. Our new method guarantees that the NLPs learned contain only minimal conditions for a variable to be true in the next state. Study of the computational complexity of our new method shows that it remains equivalent to the previous version of LF1T. Using examples from the biological literature, we show through experimental results that our specialization method can compete with the previous versions of **LF1T** in practice. We provide all proofs of theorems in the appendix.

2   Background

In this section we recall some preliminaries of logic programming. We consider a first-order language and denote the Herbrand base (the set of all ground atoms) as $\mathcal{B}$. A (normal) logic program (NLP) is a set of rules of the form

$$A \leftarrow A_1 \wedge \cdots \wedge A_m \wedge \neg A_{m+1} \wedge \cdots \wedge \neg A_n \qquad (1)$$

where A and A_i's are atoms ($n \geq m \geq 0$). For any rule R of the form (1), the atom A is called the head of R and is denoted as $h(R)$, and the conjunction to the right of $\leftarrow$ is called the body of R. We represent the set of literals in the body of R of the form (1) as $b(R) = \{A_1, \ldots, A_m, \neg A_{m+1}, \ldots, \neg A_n\}$, and the atoms appearing in the body of R positively and negatively as $b^+(R) = \{A_1, \ldots, A_m\}$ and $b^-(R) = \{A_{m+1}, \ldots, A_n\}$, respectively. The set of ground instances of all rules in a logic program P is denoted as $ground(P)$.

The Herbrand Base of a program P, denoted by $\mathcal{B}$, is the set of all atoms in the language of P. An interpretation is a subset of $\mathcal{B}$. If an interpretation is the empty set, it is denoted by ϵ. An interpretation I is a model of a program P if $b^+(R) \subseteq I$ and $b^-(R) \cap I = \emptyset$ imply $h(R) \in I$ for every rule R in P. For a logic program P and an Herbrand interpretation I, the immediate consequence operator (or T_P operator) [1] is the mapping $T_P : 2^{\mathcal{B}} \rightarrow 2^{\mathcal{B}}$:

$$T_P(I) = \{ h(R) \mid R \in ground(P), \, b^+(R) \subseteq I, \, b^-(R) \cap I = \emptyset \}. \qquad (2)$$

In the rest of this paper, we only consider rules of the form (1). To simplify the discussion we will just use the term rule.

Definition 1 (Subsumption). Let R_1 and R_2 be two rules. If $h(R_1) = h(R_2)$ and $b(R_1) \subseteq b(R_2)$ then R_1 **subsumes** R_2. If $b(R_1) \subset b(R_2)$ then R_1 is **more general** than R_2 and R_2 is **more specific** than R_1. Let S be a set of rules and R be a rule. If there exists a rule $R' \in S$ that subsumes R then S **subsumes** R. This also holds for normal logic program since a NLP is a set of rules.

In ILP, search mainly relies on generalization and specialization that are dual notions. Generalization is usually considered an induction operation, and specialization a deduction operation. In [13], the author define the minimality and maximality of the generalization and specialization operations as follows.

Definition 2 (Generalization operator [13]). A generalization operator maps a conjunction of clauses S onto a set of minimal generalizations of S. A **minimal generalization** G of S is a generalization of S such that S is not a generalization of G, and there is no generalization G' of S such that G is a generalization of G'.

Definition 3 (Specialization operator [13]). A specialization operator maps a conjunction of clauses G onto a set of maximal specializations of G. A **maximal specialization** S of G is a specialization of G such that G is not a specialization of S, and there is no specialization S' of G such that S is a specialization of S'.

The body of a rule of the form (1) can be considered as a clause, so that, the Definitions 2 and 3 can also be used to compare the body of two rules.

3   Learning from 1-Step Transitions

LF1T is an *any time algorithm* that takes a set of one-step state transitions E as input. These one-step state transitions can be regarded as positive examples. From these transitions the algorithm learns a logic program P that represents the dynamics of E. To perform this learning process we can iteratively consider one-step transitions. Figure 1 represents a Boolean nework with its corresponding state transition diagram. Given this state transition diagram as input, **LF1T** can learn the Boolean network N_1.

In *LF1T*, the Herbrand base $\mathcal{B}$ is assumed to be finite. In the input E, a state transition is represented by a pair of Herbrand interpretations. The output of *LF1T* is an NLP that realizes all state transitions of E.

Learning from 1-Step Transitions (LF1T)

Input: $E \subseteq 2^{\mathcal{B}} \times 2^{\mathcal{B}}$: (positive) examples/observations
Output: An NLP P such that $J = T_P(I)$ holds for any $(I, J) \in E$.

To construct an NLP with **LF1T** we use a bottom-up method that generates hypotheses by *generalization* from the most specific clauses or examples until every positive example is covered. **LF1T** first constructs the most specific rule R_A^I for each positive literal A appearing in $J = T_P(I)$ for each $(I, J) \in E$:

$$R_A^I := (A \leftarrow \bigwedge_{B_i \in I} B_i \land \bigwedge_{C_j \in \mathcal{B} \setminus I} \neg C_j)$$

It is important here that *we do not construct any rule to make a literal false*. For instance, from the state transition (qr, pr) **LF1T** will learn only two rules: $p \leftarrow \neg p \land q \land r$ and $r \leftarrow \neg p \land q \land r$. The rule R_A^I is then possibly generalized when another transition from E makes A true, which is computed by several generalization methods.

The two generalization methods considered in [8] are based on *resolution*. In [8], naïve and ground resolutions are defined between two ground rules as follows. Let R_1, R_2 be two ground rules and l be a literal such that $h(R_1) = h(R_2)$,

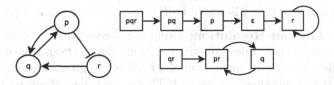

Fig. 1. A Boolean Network N_1(left) and its state transition diagram (right)

$l \in b(R_1)$ and $\bar{l} \in b(R_2)$. If $(b(R_2) \setminus \{\bar{l}\}) \subseteq (b(R_1) \setminus \{l\})$ then the *ground resolution of R_1 and R_2 (upon l)* is defined as

$$res(R_1, R_2) = \left(h(R_1) \leftarrow \bigwedge_{L_i \in b(R_1) \setminus \{l\}} L_i \right). \tag{3}$$

In particular, if $(b(R_2) \setminus \{\bar{l}\}) = (b(R_1) \setminus \{l\})$ then the ground resolution is called the *naïve resolution of R_1 and R_2 (upon l)*. In this particular case, the rules R_1 and R_2 are said to be *complementary* to each other *with respect to l*.

Both naïve resolution and ground resolution can be used as generalization methods of ground rules. For two ground rules R_1 and R_2, the naïve resolution $res(R_1, R_2)$ subsumes both R_1 and R_2, but the non-naïve ground resolution subsumes R_1 only.

Definition 4 (Consistency). *Let R be a rule and E be a set of state transitions. R is **consistent** with E if $\forall (I, J) \in E$ when $b(R) \subseteq I$ then $h(R) \in J$. Let P be a NLP. P is **consistent** with E if $\forall (I, J) \in E$, $\forall R' \in P$, when $b(R') \subseteq I$ then $h(R') \in J$.*

Ground and naïve resolutions can be used to learn a ground NLP. Both methods keep the consistency of the learned rules. For example, let us consider the three rules: $R_1 = (p \leftarrow q \wedge r)$, $R_2 = (p \leftarrow \neg q \wedge r)$, $R_3 = (p \leftarrow \neg q)$, and their resolvent: $res(R_1, R_2) = res(R_1, R_3) = (p \leftarrow r)$. R_1 and R_2 are complementary with respect to q. Both R_1 and R_2 can be generalized by the naïve resolution of them because $res(R_1, R_2)$ subsumes both R_1 and R_2. On the other hand, the ground resolution $res(R_1, R_3)$ of R_1 and R_3 is equivalent to $res(R_1, R_2)$. However, $res(R_1, R_3)$ subsumes R_1 but does not subsume R_3.

LF1T with Naïve Resolution: In the first implementation of **LF1T** of [8], naïve resolution is used as a least generalization method. This method is particularly intuitive from the ILP viewpoint, since each generalization is performed based on a least generalization operator. In [8], it is shown that for two complementary ground rules R_1 and R_2, the naïve resolution of R_1 and R_2 is the least generalization [15] of them, that is, $lg(R_1, R_2) = res(R_1, R_2)$. When naïve resolution is used, **LF1T** needs an auxiliary set P_{old} of rules to globally store subsumed rules, which increases monotonically. P_{old} is set to be $\emptyset$ at first. When a generated rule R is newly added **LF1T** searches a rule $R' \in P \cup P_{old}$ such that $h(R') = h(R)$ and $b(R)$ and $b(R')$ differ in the sign of only one literal l. If there is no such a rule R', then R is just added to P; otherwise, R and R' are added to P_{old} and then $res(R, R')$ is added to P.

LF1T with Ground Resolution: Using naïve resolution, $P \cup P_{old}$ possibly contains all patterns of rules constructed from the Herbrand base $\mathcal{B}$ in their bodies. In the second implementation of **LF1T** of [8], ground resolution is used as an alternative generalization method. This replacement of resolution leads to a lot of computational gains since the use of P_{old} is not necessary any more: all generalized rules obtained from $P \cup P_{old}$ by naïve resolution can be obtained using ground resolution on P. By Theorem 3 of [8], using the naïve version, the memory use of the **LF1T** algorithm is bounded by $O(n \cdot 3^n)$, and the time complexity of learning is bounded by $O(n^2 \cdot 9^n)$, where $n = |\mathcal{B}|$. On the other hand, with ground resolution, the memory use is bounded by $O(2^n)$, which is the maximum size of P, and the time complexity is bounded by $O(4^n)$. Given the set E of complete state transitions, which has the size $O(2^n)$, the complexity of **LF1T**$(E, \emptyset)$ with ground resolution is bounded by $O(|E|^2)$. On the other hand, the worst-case complexity of learning with naïve resolution is $O(n^2 \cdot |E|^{4.5})$.

4   Learning Prime Implicant Conditions

In this section, we use the notion of prime implicant to define minimality of NLP. We consider that the NLP learn by **LF1T** is minimal if the body of each rule constitutes a prime implicant to infer the head.

Definition 5 *(Prime Implicant Condition). Let R be a rule and E a set of state transitions such that R is consistent with E. $b(R)$ is a **prime implicant condition** of $h(R)$ for E if there does not exist another rule R' consistent with E such that R' subsumes R. Let P be a NLP such that $P \cup \{R\} \equiv P$: all models of $P \cup \{R\}$ are models of P and vice versa. $b(R)$ is a **prime implicant condition** of $h(R)$ for P if there does not exist another rule R' such that $P \cup \{R'\} \equiv P$ and R' subsumes R.*

Definition 6 *(Prime Rule). Let R be a rule and E a set of state transitions such that R is consistent with E. Let P be a NLP such that $P \cup \{R\} \equiv P$. R is a **prime rule** of E (resp. P) if $b(R)$ is a prime implicant condition of $h(R)$ for E (resp. P). For any atom p the **most general prime rule** for p is the rule with an empty body $(p \leftarrow)$ that states that p is always true in the next state.*

Example 1. Let R_1, R_2 and R_3 be three rules and E be the set of state transitions of Fig. 1 as follows: $R_1 = p \leftarrow p \wedge q \wedge r$, $R_2 = p \leftarrow p \wedge q$, $R_3 = p \leftarrow q$ The only rule more general than R_3 is $R' = p.$, but R' is not consistent with $(p, \epsilon) \in E$ so that R_3 is a prime rule for E. Since R_3 subsumes both R_1 and R_2, they are not prime rules of E. Let P be the NLP $\{p \leftarrow p, q \leftarrow p \wedge r, r \leftarrow \neg p\}$, R_3 is a prime rule of P because P realizes E and R_3 is minimal for E.

Definition 7 *(Prime NLP). Let P be an NLP and E be the state transitions of P, P is a **prime NLP** for E if P realizes E and all rules of P are prime rule for E. We call the set of all prime rules of E the **complete prime NLP** of E.*

Example 2. Let R_1, R_2 and R_3 be three rules, E be the set of state transitions of Fig. 1 and P an NLP as follows: $R_1 = p \leftarrow p \wedge q$, $R_2 = q \leftarrow p \wedge r$, $R_3 = r \leftarrow \neg p$ and $P = \{R_1\} \cup \{R_2\} \cup \{R_3\}$. Since R_1, R_2 and R_3 are prime rule for E, P the NLP formed of these three rules is a prime NLP of E. There does not exist any other prime rules for E, therefore P is also the complete prime NLP of E.

The complete prime NLP of a given set of state transitions E can naïvely be obtained by brute force search. Starting from the most general rules, that is, fact rules, it suffices to generate all maximal specific specialization step by step and keep the first ones that are consistent with E. This method implies to check all state transitions for all possible rules that correspond to $O(n \times 3^n \times 2^n) = O(6^n)$ checking operations in the worst case for a Herbrand base of n variables. But it is also possible to do it by extending previous **LF1T** algorithm for the sake of complexity. Here we propose a simple extension of naïve (resp. ground) resolution. In previous algorithms, for each rule learned, only the first least generalization found is kept. Now we consider all possible least generalizations and define full naïve (resp. ground) resolution. **LF1T** with full naïve (resp. ground) resolution learn the complete prime NLP that realize the input state transitions.

Definition 8 *(full naive resolution and full ground resolution). Let R be a rule and P be a NLP. Let P_R be a set of rule of P such that, for all $R' \in P_R$, $h(R) = h(R')$ and for each R' there exists $l \in b(R)$, $(b(R') \setminus \{\bar{l}\}) = (b(R) \setminus \{l\})$ (resp. $(b(R') \setminus \{\bar{l}\}) \subseteq (b(R) \setminus \{l\})$). The* **full naïve (resp. ground) resolution** *of R by P is the set of all possible naïve (resp. ground) resolutions of R with the rules of P: $res_f(R, P) = \{res(R, R') | R' \in P_R\}$.*

Theorem 1 *(**Completeness and Soundness of full resolution**). Given a set E of pairs of interpretations, **LF1T** with full naïve (resp. ground) resolution is complete and sound for E.*

Theorem 2 *(**LF1T with full resolution learns complete prime NLP**). Given a set E of pairs of interpretations, **LF1T** with full naïve (resp. ground) resolution will learn the complete prime NLP that realizes E.*

4.1   Least Specialization for LF1T

Until now, to construct an NLP, **LF1T** relied on a bottom-up method that generates hypotheses by *generalization* from the most specific clauses or examples until every positive example is covered. This time we propose a new learning method that generate hypotheses by *specialization* from the most general rules until no negative example is covered. Learning by specialization ensures to output the most general valid hypothesis. That is similar to the notion of specialization we use here amoung body of rules with the same head. In ILP, refinement operators usually apply a substitution θ and add a set of literals to a clause [13]. Similarly, in our new algorithm, we refine rules by adding the negation of negative transitions into their body.

Definition 9 *(Least specialization). We call a* **maximal specialization** *of a rule R, a rule R_S if $h(R_S) = h(R)$ and $b(R_S)$ is a maximal specialization of $b(R)$. Let R_1 and R_2 be two rules such that $h(R_1) = h(R_2)$ and R_1 subsumes R_2, e.g. $b(R_1) \subseteq b(R_2)$. Let l_i be the ith literal of $b(R_2)$, then the* **least specialization** *of R_1 over R_2 is as follows:*

$$ls(R_1, R_2) = \{h(R_1) \leftarrow (b(R_1) \land \neg b(R_2))\} = \{(h(R_1) \leftarrow (b(R_1) \land \bar{l_i}) | l_i \in b(R_2) \setminus b(R_1)\}$$

Let P be an NLP, R be a rule and S be the set of all rules of P that subsumes R. The **least specialization** *$ls(P, R)$ of P by R is as follow*

$$ls(P, R) = (P \backslash S) \cup (\bigcup_{R_P \in S} ls(R_P, R))$$

The least specialization of a rule R can be used to avoid the subsumption of another rule with a minimal reduction of the generality of R. The least specialization of an NLP P can be used to avoid the coverage of a negative transition with a minimal reduction of the generality of the rules of P.

Theorem 3 *(Soundness of least specialization).* *Let R_1, R_2 be two rules such that R_1 subsumes R_2. Let S_1 be the set of rules subsumed by R_1 and S_2 be the rules of S_1 that subsume R_2. The least specialization of R_1 by R_2 only subsumes the set of rules $S_1 \backslash S_2$. Let P be a NLP and R be a rule such that P subsumes R. Let S_P be the set of rules subsumed by P and S_R be the rules of S_P that subsume R. The least specialization of P by R only subsumes the set of rules $S_P \backslash S_R$.*

5   Algorithm

Now we present a new **LF1T** algorithm based on least specialization. The novelty of this approach is double: first it relies on specialization in place of generalization and most importantly, it guarantees that the output is the complete prime NLP that realize the input transitions, as shown by Theorem 5. Algorithm 1 shows the pseudo-code of **LF1T** with least specialization. Like in previous versions, **LF1T** takes a set of state transitions E as input and outputs an NLP P that realizes E. To guarantee the minimality of the learned NLP, **LF1T** starts with an initial NLP $P_0^{\mathcal{B}}$ that is the most general complete prime NLP of the Herbrand base $\mathcal{B}$ of E, i.e. the NLP that contains only facts (lines 3–7): $P_0^{\mathcal{B}} = \{p. | p \in \mathcal{B}\}$. Then **LF1T** iteratively analyzes each transition $(I, J) \in E$ (lines 8–13).

For each variable A that **does not appear** in J, **LF1T** infers an **anti-rule** R_A^I (lines 11–12):

$$R_A^I := A \leftarrow \bigwedge_{B_i \in I} B_i \wedge \bigwedge_{C_j \in (\mathcal{B} \backslash I)} \neg C_j$$

A is in the head as it denotes a negative example. Then, **LF1T** uses least specialization to make P consistent with all R_A^I (line 12). Algorithm 2 shows in detail the pseudo code of this operation. **LF1T** first extracts all rules $R_P \in P$ that subsume R_A^I (lines 3–8). It generates the least specialization of each R_P by generating a rule for each literal in R_A^I (lines 9–12). Each rule contains all literals of R_P plus the opposite of a literal in R_A^I so that R_A^I is not subsumed by that rule. Then **LF1T** adds in P all the generated rules that are not subsumed by P (line 13–15), so that P becomes consistent with the transition (I, J) and remains a complete prime NLP. When all transitions have been analyzed, **LF1T** outputs P that has become the complete prime NLP of E.

Algorithm 1. LF1T(E) : Learn the complete prime NLP P of E

1: INPUT: $\mathcal{B}$ a set of atoms and $E \subseteq 2^{\mathcal{B}} \times 2^{\mathcal{B}}$
2: OUTPUT: An NLP P such that $J = T_P(I)$ holds for any $(I, J) \in E$.

3: P a NLP
4: $P := \emptyset$
    // Initialize P with the most general rules
5: **for** each $A \in \mathcal{B}$ **do**
6:     $P := P \cup \{A.\}$
7: **end for**
    // Specify P by interpretation of transitions
8: **while** $E \neq \emptyset$ **do**
9:     Pick $(I, J) \in E$; $E := E \setminus \{(I, J)\}$
10:     **for** each $A \in \mathcal{B}$ **do**
11:         **if** $A \notin J$ **then**
12:             $R_A^I := A \leftarrow \bigwedge_{B_i \in I} B_i \wedge \bigwedge_{C_j \in (\mathcal{B} \setminus I)} \neg C_j$
13:             $P := \textbf{Specialize}(P, R_A^I)$
14:         **end if**
15:     **end for**
16: **end while**
17: **return** P

Algorithm 2. specialize(P,R) : specify the NLP P to not subsume the rule R

1: INPUT: an NLP P and a rule R
2: OUTPUT: the maximal specific specialization of P that does not subsumes R.

3: $conflicts$: a set of rules
4: $conflicts := \emptyset$
    // Search rules that need to be specialized
5: **for** each rule $R_P \in P$ **do**
6:     **if** $b(R_P) \subseteq b(R)$ **then**
7:         $conflicts := conflicts \cup R_P$
8:         $P := P \setminus R_P$
9:     **end if**
10: **end for**
    // Revise the rules by least specialization
11: **for** each rule $R_c \in conflicts$ **do**
12:     **for** each literal $l \in b(R)$ **do**
13:         **if** $l \notin b(R_c)$ and $\bar{l} \notin b(R_c)$ **then**
14:             $R_c' := (h(R_c) \leftarrow (b(R_c) \cup \bar{l}))$
15:             **if** P does not subsumes R_c' **then**
16:                 $P := P \setminus$ all rules subsumed by R_c'
17:                 $P := P \cup R_c'$
18:             **end if**
19:         **end if**
20:     **end for**
21: **end for**
22: **return** P

Table 1. Execution of **LF1T** with least specialization on state transitions of Fig. 1

Initialization	$pqr \to pq$	$pq \to p$	$p \to \epsilon$	$\epsilon \to r$	$r \to r$
$p.$	$p.$	$p.$	~~$p \leftarrow \neg p.$~~	$p \leftarrow q.$	$p \leftarrow q.$
$q.$	$q.$	$q \leftarrow \neg p.$	$p \leftarrow q.$	$p \leftarrow r.$	$p \leftarrow p \wedge r.$
$r.$	$r \leftarrow \neg p.$	$q \leftarrow \neg q.$	$p \leftarrow r.$	~~$p \leftarrow \neg p \wedge q.$~~	~~$p \leftarrow q \wedge r.$~~
	$r \leftarrow \neg q.$	$q \leftarrow r.$	$q \leftarrow \neg p.$	~~$p \leftarrow \neg p \wedge r.$~~	$q \leftarrow \neg p \wedge q.$
	$r \leftarrow \neg r.$	$r \leftarrow \neg p.$	$q \leftarrow r.$	~~$q \leftarrow \neg p \wedge q.$~~	$q \leftarrow p \wedge r.$
		$r \leftarrow \neg q.$	~~$q \leftarrow \neg p \wedge \neg q.$~~	~~$q \leftarrow \neg p \wedge r.$~~	$q \leftarrow q \wedge r.$
		~~$r \leftarrow \neg p \wedge \neg r.$~~	$q \leftarrow \neg p \wedge q.$	$q \leftarrow \neg p \wedge r.$	$r \leftarrow \neg p.$
		~~$r \leftarrow \neg q \wedge \neg r.$~~	$r \leftarrow \neg p.$	$r \leftarrow \neg p.$	$r \leftarrow \neg q \wedge r.$
			~~$r \leftarrow \neg p \wedge \neg q.$~~	$r \leftarrow \neg q \wedge r.$	
			$r \leftarrow \neg q \wedge r.$		

$qr \to pr$	$pr \to q$	$q \to pr$
$p \leftarrow q.$	$p \leftarrow q.$	$p \leftarrow q.$
$p \leftarrow p \wedge r.$	~~$p \leftarrow p \wedge q \wedge r.$~~	$q \leftarrow p \wedge r.$
$q \leftarrow p \wedge r.$	$q \leftarrow p \wedge r.$	$r \leftarrow \neg p.$
$q \leftarrow \neg p \wedge q \wedge \neg r.$	$q \leftarrow \neg p \wedge q \wedge \neg r.$	
~~$q \leftarrow p \wedge q \wedge r.$~~	$r \leftarrow \neg p.$	$q \leftarrow \neg p \wedge q \wedge \neg r.$
$r \leftarrow \neg p.$	~~$r \leftarrow \neg p \wedge \neg q \wedge r.$~~	is removed because
$r \leftarrow \neg q \wedge r.$		it cannot be specialized

Table 1 shows the execution of **LF1T** with least specialization on step transitions of Fig. 1 where $pqr \to pq$ represents the state transition $(\{p,q,r\},\{p,q\})$. Introduction of literal by least specialization is represented in bold and rules that are subsumed after specialization are stroked. **LF1T** starts with the most general set of prime rules that can realize E, that is $P = \{p.,q.,r.\}$. From the transition (pqr,pq) **LF1T** infers the rule $r \leftarrow p \wedge q \wedge r$ that is subsumed by $r. \in P$. **LF1T** then replaces that rule by its least specialization: $ls(r, r \leftarrow p \wedge q \wedge r) = \{r \leftarrow \neg p, r \leftarrow \neg q, r \leftarrow \neg r\}$ Furthermore, P becomes consistent with (pqr,pq). From (pq,p) **LF1T** infers two rules: $q \leftarrow p \wedge q \wedge \neg r$ and $r \leftarrow p \wedge q \wedge \neg r$, that are respectively subsumed by $q.$ and $r \leftarrow \neg r$. The first rule, $q.$, is replaced by its least specialization: $\{q \leftarrow \neg p, q \leftarrow \neg q, q \leftarrow r\}$. For the second rule, $r.$, its least specialization by $r \leftarrow p \wedge q \wedge \neg r$ generates two rules, $r \leftarrow \neg p \wedge \neg r$ and $r \leftarrow \neg q \wedge \neg r$. But these rules are respectively subsumed by $r \leftarrow \neg p$ and $r \leftarrow \neg q$ that are already in P. The subsumed rules are not added to P, so that the analysis of (pq,p) results in the specialization of $q.$ and the deletion of $r \leftarrow \neg r$.

Learning continues with similar cases until the last transition (q,pr) where we have a special case. From this transition, **LF1T** infers the rule $q \leftarrow \neg p \wedge q \wedge \neg r$ that is subsumed by P on $R := q \leftarrow \neg p \wedge q \wedge \neg r$. Because $|b(R)| = |\mathcal{B}|$ it cannot be specialized so that P becomes consistent with (q,pr), **LF1T** just removes R from P.

Theorem 4 (Completeness of LF1T with least specialization). *Let $P_0^{\mathcal{B}}$ be the most general complete prime NLP of a given Herbrand base $\mathcal{B}$. Initializing*

LF1T with $P_0^\mathcal{B}$, by using least specialization iteratively on a set of state transitions E, **LF1T** learns an NLP that realizes E.

Theorem 5 (LF1T with least specialization outputs a complete prime NLP). Let $P_0^\mathcal{B}$ be the most general complete prime NLP of a given Herbrand base $\mathcal{B}$. Initializing **LF1T** with $P_0^\mathcal{B}$, by using least specialization iteratively on a set of state transitions E, **LF1T** learns the complete prime NLP of E.

Theorem 6 (Complexity). Let n be the size of the Herbrand base $|\mathcal{B}|$. Using least specialization, the memory complexity of **LF1T** remains in the same order as the previous algorithms based on ground resolution, i.e., $O(2^n)$. But the computational complexity of **LF1T** with least specialization is higher than the previous algorithms based on ground resolution, i.e. $O(n \cdot 4^n)$ and $O(4^n)$, respectively. Same complexity results for full naïve (resp. ground) resolution.

6 Evaluation

In this section, we evaluate our new learning methods through experiments. We apply our new **LF1T** algorithms to learn Boolean networks. Here we run our learning program on the same benchmarks used in [8,16]. These benchmarks are Boolean networks taken from Dubrova and Teslenko [3], which include those networks about control of flower morphogenesis in *Arabidopsis thaliana*, budding yeast cell cycle regulation, fission yeast cell cycle regulation, mammalian cell cycle regulation and T helper cell cycle regulation. Like in [8,16], we first construct an NLP $\tau(N)$ from the Boolean function of a Boolean network N where each Boolean function is transformed into a DNF formula. Then, we get all possible 1-step state transitions of N from all $2^{|\mathcal{B}|}$ possible initial states I^0's by computing all stable models of $\tau(N) \cup I^0$ using the answer set solver clasp [5]. Finally, we use this set of state transitions to learn an NLP using our **LF1T** algorithm. Because a run of **LF1T** returns an NLP which can contain redundant rules, the original NLP P_{org} and the output NLP P_{LFIT} of **LF1T** can be different, but remain equivalent with respect to state transition, that is, $T_{P_{org}}$ and $T_{P_{LFIT}}$ are identical functions. Regarding the new algorithms, it can also be the case if the original NLP is not a complete prime NLP. For the new versions of **LF1T**, if P_{org} is not a prime complete NLP we will learn a simplification of P_{org}. Table 2 shows the memory space and time of a single **LF1T** run in learning a Boolean network for each benchmark on a processor Intel Core I7 (3610QM, 2.3 GHz) with 4 GB of RAM. It compares memory and run time of the three previous algorithm (naïve, ground and the BDD optimization of the ground version) with their extension to learn complete prime NLP and the new algorithm based on least specialization. For each version of **LF1T** the variable ordering is alphabetical and transition ordering is the one that clasp outputs. The time limit is set to two hours for each experiment. Memory is represented in (maximal) number of literal in the NLP learned. Except for LF1T-BDD, all implemented algorithms uses the same data structures. That is why even **LF1T** with least specialization cannot compete with the ground-BDD version regarding memory

Table 2. Memory use and learning time of **LF1T** for Boolean networks benchmarks up to 23 nodes in the same condition as in [8]

Algorithm	Mammalian (10)	Fission (10)	Budding (12)	Arabidopsis (16)	T helper (23)
Naïve	142 118/4.62 s	126 237/3.65 s	1 147 124/523 s	T.O	T.O.
Ground	1036/0.04 s	1218/0.05 s	21 470/0.26 s	271 288/4.25 s	T.O.
Ground-BDD	180/0.24 s	147/0.24 s	**541/0.19 s**	**779/2.8 s**	611/3360 s
Full naïve	377 539/29.25 s	345587/24.03 s	T.O	T.O	T.O.
Full ground	1066/0.24 s	1178/0.23 s	23 738/4.04 s	399 469/111 s	T.O.
Least specialization	**375/0.06 s**	**377/0.08 s**	641/0.35 s	2270/5.28 s	3134/5263 s

and run time. It is more relevant to compare it to the original implementation of **LF1T** with ground resolution and the new one with full ground resolution.

On Table 2 we can observe that, as the number of variable increases, the memory efficiency of least specialization regarding ground version becomes more interesting. Regarding run time, both algorithms have globally equivalent performances. But least specialization ensure that the output is unique in the fact that it is the complete prime NLP of the given input transitions. **LF1T** with full ground resolution also ensure this property, but is much less efficient than least specialization regarding both memory use and run time. On the benchmark, least specialization is respectively 75 %, 65 %, 91 % and 95 % faster. Least specialization version also succeed to learn the t-helper benchmark (23 variables) in 1 h and 21 min. The main interest of using least specialization is that it guarantees to obtain a unique NLP that contains all minimal conditions to make a variable true. Previous versions of **LF1T** do not have this property and experimental results showed that their output is sensitive to variable ordering and especially transition ordering. For a given set of state transitions E, the output of **LF1T** with least specialization is always the same whatever the variable ordering or transition ordering. It is easy to see that variable ordering has no impact on both learning time and memory use of the new versions of **LF1T** since they consider all generalizations/specializations. But transition ordering has a significant impact on the learning time of the new version of **LF1T** compared to previous ones. On all experiments we run, the ordering of the output of clingo gives the best results. More investigation are required to determine if we can design a heuristic to make a good ordering of the input transition to speed up the run time of the new algorithms.

7   Conclusion and Future Work

We proposed a new algorithm for learning from interpretation transitions based on least specialization. Given any state transition diagram we can learn an NLP that exactly captures the system dynamics. Learning is performed only from positive examples, and produces NLPs that consist only of rules to make literals true. Consistency of state transition rules is achieved by least specialization, in which minimality of rules is guaranteed. As a result, given any state transition diagram E, **LF1T** with least specialization always learns a unique NLP that contains all prime rules that realize E. It implies that the output of **LF1T** is no more sensitive to variable ordering or transition ordering. But, experimental results showed that the new algorithm is sensitive to input transitions ordering regarding run time. Design of an heuristic to make a good ordering of the input is one possible future work. We are now considering to extend our framework to learn non-deterministic dynamic systems. One of our expectation is to be able to learn probabilistic logic progr am from interpretation of transitions. Assuming that probability of transition are given as input it should be possible to infer probabilistic rules using adapted LFIT techniques. But how to combine probabilities when generalization/specialization occurs is an interesting problem that we plan to tackle in our future works.

A   Appendix

A.1   Proof of Theorem 1 (Completeness)

Given a set E of pairs of interpretations, **LF1T** with full naïve (resp. ground) resolution is complete for E.

Proof. According to Theorem 1 (resp. 2) of [8], **LF1T** with naïve (resp. ground) resolution is complete for E. It is trivial that any rules produced by naïve (resp. ground) resolution can be obtained by full naïve (resp. ground) resolution. Then, if P and P' are respectively obtained by naïve (resp. ground) resolution and full naïve (resp. ground) resolution, P' theory-subsumes P. If a program P is complete for E , a program P' that theory-subsumes P is also complete for E. Since P is complete for E by Theorem 1 of [8], P' is complete for E. □

A.2   Proof of Theorem 1: (Soundness)

Given a set E of pairs of interpretations, **LF1T** with full naïve (resp. ground) resolution is sound for E.

Proof. All rules that can be produced by naïve (resp. ground) resolution can be obtained by full naïve (resp. ground) resolution. Since all rules produced by naïve (resp. ground) resolution are sound for E (Corrollary 1 (resp. 2) of [8]), full naïve (resp. ground) resolution is sound for E. □

A.3   Proof of Theorem 2

Given a set E of pairs of interpretations, **LF1T** with full naïve (resp. ground) resolution learn the complete prime NLP that realize E.

Proof. Let us assume that **LF1T** with full naïve resolution does not learn a prime NLP of E. If our assumption is correct it implies that there exists R a prime rule for E that cannot be learned by **LF1T** with full naïve resolution. Let $\mathcal{B}$ be the herbrand base of E.

Case 1: $|b(R)| = |\mathcal{B}|$, R will be directly infer from a transition $(I, J) \in E$. This is a contradiction with our assumption.

Case 2: $|b(R)| < |\mathcal{B}|$. let l be a literal such that $l \notin b(R)$, according to our assumption, their is a rule R' that is one of the rule $R_1 := h(R) \leftarrow b(R) \cup l$ or $R_2 := h(R) \leftarrow b(R) \cup \bar{l}$ and R' cannot be learned because $res(R_1, R_2) = R$. Recursively, what applies to R applies to R' until we reach a rule R'' such that $|b(R'')| = |\mathcal{B}|$. Our assumption implies that this rule R'' cannot be learned, but R'' will be directly infer from a transition $(I, J) \in E$, this is a contradiction. Since ground resolution can learn all rules learn by naïve resolution, the proof also applies to **LF1T** with full ground resolution.$\square$

A.4   Proof of Theorem 3

Let R_1, R_2 be two rules such that $b(R_1) \subseteq b(R_2)$. Let S_1 be the set of rules subsumed by R_1 and S_2 be the rules of S_1 that subsume R_2. The least specialization of R_1 by R_2 only subsumes the set of rules $S_1 \backslash S_2$.

Proof. : According to Definition 9, the least specialization of R_1 by R_2 is as follows:
$$ls(R_1, R_2) = \{h(R_1) \leftarrow (b(R_1) \wedge \neg b(R_2))\}$$

All rule R of S_2 subsumes R_2, then according to Definition 1 $b(R) \subseteq b(R_2)$. If $ls(R_1, R_2)$ subsumes an R then there exists $R' \in ls(R_1, R_2)$ and $b(R') \subseteq b(R)$. Since $R' \in ls(R_1, R_2)$, there is a $l \in b(R_2)$ such that $\bar{l} \in b(R')$, so that $b(R') \not\subseteq b(R_2)$. Since all $R \in S_2$ subsume R_2, R' cannot subsume any R since R' does not subsume R_2.

Conclusion 1: the least specialization of R_1 by R_2 cannot subsume any $R \in S_2$.

Let us suppose there is a rule $R' \in S_1$ that does not subsumes R_2 and is not subsumed by $ls(R_1, R_2)$. Let l_i be the ith literal of $b(R_2)$, then:

$$ls(R_1, R_2) = \{(h(R_1) \leftarrow (b(R_1) \wedge \bar{l_i})|l_i \in b(R_2) \backslash b(R_1)\} (1)$$

R' is subsumed by R_1, so that $R' = h(R_1) \leftarrow b(R_1) \cup S$, with S a set of literal. R' does not subsume R_2, so that there exists a $l \in b(R_2) \backslash b(R_1)$ such that $\bar{l} \in S$. According to (1), the rule $R'' = h(R_1) \leftarrow b(R_1) \wedge \bar{l}$ is in $ls(R_1, R_2)$. Since R'' subsumes R' and $R'' \in ls(R_1, R_2)$, $ls(R_1, R_2)$ subsumes R'.

Conclusion 2: the least specialization of R_1 by R_2 subsumes all rule of S_1 that does not subsume R_2.

Final conclusion: the least specialization of R_1 by R_2 only subsumes $S_1 \backslash S_2$.$\square$

Now, let P be an NLP and R be a rule such that P subsumes R. Let S_P be the set of rules subsumed by P and S_R be the rules of S_P that subsume R. The least specialization of P by R only subsumes the set of rules $S_P \backslash S_R$.

Proof. : According to Definition 5, the least specialization $ls(P, R)$ of P by R is as follows:
$$ls(P, R) = (P \backslash S_P) \cup (\bigcup_{R_P \in S_P} ls(R_P, R))$$

For any rule R_P let S_{R_P} be the set of rules subsumed by R_P and $S_{R_P2} \in S_R$ be the rule of S_{R_P} that subsume R.

According to Theorem 3 the least specialization of R_P by R only subsumes $S_{R_P} \backslash S_{R_P2}$. So that $\bigcup\limits_{R_P \in S_P} ls(R_P, R)$ only subsumes ($\bigcup\limits_{R_P \in S_P} S_{R_P} \backslash S_{R_P2}) = $ ($\bigcup\limits_{R_P \in S_P} S_{R_P}) \backslash S_R$. Then $ls(P, R)$ only subsumes the rules subsumed by $(P \backslash S_P) \cup$ ($\bigcup\limits_{R_P \in S_P} S_{R_P}) \backslash S_R$, that is $S_P \backslash S_R$.

Conclusion: The least specialization of P by R only subsumes $S_P \backslash S_R$. □

A.5   Proof of Theorem 4

Let $P_0^{\mathcal{B}}$ be the most general complete prime NLP of a given Herbrand base $\mathcal{B}$, i.e. the NLP that contains only facts

$$P_0^{\mathcal{B}} = \{p.|p \in \mathcal{B}\}$$

Initializing **LF1T** with $P_0^{\mathcal{B}}$, by using least specialization iteratively on the transitions of a set of state transitions E, **LF1T** learns an NLP P that realizes E.

Proof. : Let P be an NLP consistent with a set of transitions E', S_P be the set of rules subsumed by P and a state transition (I, J) such that $E' \subset E$ and $(I, J) \in E$ but $(I, J) \notin E'$. According to Theorem 3, for any rule R_A^I that can be inferred by **LF1T** from (I, J) that is subsumed by P, the least specialization $ls(P, R_A^I)$ of P by R_A^I exactly subsumes the rules subsumed by P except the ones subsumed by R_A^I. Since $|R_A^I|$ is $|\mathcal{B}|$, R_A^I only subsumes itself so that $ls(P, R)$ exactly subsumes $S_P \backslash R_A^I$. Let P' be the NLP obtained by least specialization of P with all R_A^I that can be inferred from (I, J), then P' is consistent with $E' \cup \{(I, J)\}$.

Conclusion 1: LF1T keep the consistency of the NLP learned.

LF1T start with $P_0^{\mathcal{B}}$ as initial NLP. $P_0^{\mathcal{B}}$ is at least consistent with $\emptyset \subseteq E$. According to conclusion 1, initializing **LF1T** with $P_0^{\mathcal{B}}$ and by using least specialization iteratively on the element of E when its needed, **LF1T** learns an NLP that realizes E.□

A.6   Proof of Theorem 5

Let $P_0^{\mathcal{B}}$ be the most general complete prime NLP of a given Herbrand base $\mathcal{B}$, i.e. the NLP that contains only facts

$$P_0^{\mathcal{B}} = \{p.|p \in \mathcal{B}\}$$

Initializing **LF1T** with $P_0^{\mathcal{B}}$, by using least specialization iteratively on a set of state transitions E, **LF1T** learns the complete prime NLP of E.

Proof. : Let us assume that **LF1T** with least specialization does not learn a prime NLP of E. If our assumption is correct, according to Theorem 4, **LF1T** learns a NLP P, that is consistent with E and P is not the complete prime NLP of E. **LF1T** start with $P_0^{\mathcal{B}}$ as initial NLP, $P_0^{\mathcal{B}}$ is the most general complete prime NLP that can cover E.

Consequence 1: LF1T with least specialization can transform a complete prime NLP into an NLP that is not a complete prime NLP.

Let P be the complete prime NLP of a set of state transition $E' \subset E$ and $(I, J) \notin E'$, such that P is not consistent with (I, J). Our assumption implies that the least specialization P' of P by the rules inferred from (I, J) is not the complete prime NLP of $E' \cup (I, J)$. According to Definition 7, there is two possibilities:

- case 1: $\exists R \in P'$ such that R is not a prime rule of $E' \cup (I, J)$.
- case 2: $\exists R' \notin P'$ such that R' is a prime rule of $E' \cup (I, J)$.

Case 1.1: If $R \in P$, it implies that R is a prime rule of E' and that R is consistent with (I, J), otherwise R should have been specialized. Because R is not a prime rule of $E' \cup (I, J)$ it implies that there exists a rule R_m consistent with $E' \cup (I, J)$ that is more general than R, i.e. $b(R_m) \subset b(R)$. Then R_m is also consistent with E', but since R is a prime rule of E' there does not exist any rule consistent with E' that is more general than R. This is a contradiction.

Case 1.2: Now let us suppose that $R \notin P$; then R has been obtained by least specialization of a rule $R_P \in P$ by a rule inferred from (I, J). It implies that $\exists l \in b(R)$ and $\bar{l} \in I$. If R is not a prime rule of $E' \cup (I, J)$, there exists R_m a prime rule of $E' \cup (I, J)$ and R_m is more general than R. It implies that $l \in R_m$ otherwise R_m is not consistents with (I, J) because it will also subsumes R_P that is not consistents with (I, J). Since R_m is consistent with $E' \cup (I, J)$ it is also consistent with E'. This implies that $\exists R_m'$ a prime rule of E' that subsumes R_m (it can be R_m itself), R_m' also subsumes R. Since P is the complete prime NLP of E', $R_m' \in P$.

Case 1.2.1: Let suppose that $l \notin b(R_m')$, since $l \in b(R)$ and R_m' subsumes R then R_m' subsumes R_P because $R = h(R_P) \leftarrow b(R_P) \cup l$. But since R_P is a prime rule of E' it implies that $R_m' = R_P$. In that case it means that R_P subsumes R_m and since $l \in R_m$, $h(R_P) \leftarrow b(R_P) \cup l$ also subsumes R_m. Since $h(R_P) \leftarrow b(R_P) \cup l$ is R, R subsumes R_m and R_m can neither be more general than R nor a prime rule of $E' \cup (I, J)$. This is a contradiction with case 2.

Case 1.2.2: Finally let us suppose that $l \in b(R_m')$, since R_m' is consistent with E and $l \in I$, R_m' is consistent with $E' \cup (I, J)$. But R_m' subsumes R_m and since R_m is a prime rule of $E' \cup (I, J)$ it implies that $R_m' = R_m$. In that case $R_m \in P$ and because R_m is consistent with (I, J) and R_m subsumes R, **LF1T** will not add R into P'. This is a contradiction with case 1.

Case 2: Let consider that there exists a $R' \notin P'$ such that R' is a prime rule of $E' \cup (I, J)$. Since $R' \notin P'$, $R' \notin P$ and R' is not a prime rule of E' since P is the complete prime NLP of E'. Then, there exists $R_m \in P$ a prime rule of E' such that R_m subsumes R' and $R_m \notin P'$ since R' is a prime rule of $E' \cup (I, J)$. Then, $b(R') = b(R_m') \cup S$ with S a non-empty set of literals such that for all $l \in S, l \notin b(R_m)$. Since $R_m \notin P'$, there is a rule $R_{h(R_m)}^I$ that can be inferred from

(I, J) and subsumed by R_m. And there is no rule $R'_m \in ls(R_m, R^I_{h(R_m)})$ that subsumes R' since R' is a prime rule of $E' \cup (I, J)$. Then, for all $l' \in b(R^I_{h(R_m)})$, $\overline{l'} \notin b(R')$ otherwise there is a R'_m that subsumes R'. Since $|b(R^I_{h(R_m)})| = \mathcal{B}$, $b(R')$ cannot contains a literal that is not in $b(R^I_{h(R_m)})$ so that R' subsumes $R^I_{h(R_m)}$. R' cannot be a prime rule of $E' \cup (I, J)$ since R' is not consistent with (I, J), this is a contradiction.

Conclusion: If P is a complete prime NLP of $E' \subset E$, for any $(I, J) \in E$ **LF1T** with least specialization will learn the complete prime NLP P' of $E' \cup (I, J)$. Since **LF1T** starts with a complete prime NLP that is $P^{\mathcal{B}}_0$, according to Theorem 4, **LF1T** will learn a NLP consistent with E, our last statement implies that this NLP is the complete prime NLP of E since **LF1T** cannot specify a complete prime NLP into an NLP that is not a complete prime NLP. □

A.7 Proof of Theorem 6

Let n be the size of the Herbrand base $|\mathcal{B}|$. Using least specialization, the memory complexity of **LF1T** remains in the same order as the previous algorithms based on ground resolution, i.e., $O(2^n)$. But the computational complexity of **LF1T** with least specialization is higher than the previous algorithms based on ground resolution, i.e. $O(n \cdot 4^n)$ and $O(4^n)$, respectively. Same complexity results for full naïve (resp. ground) resolution.

Proof. Let n be the size of the Herbrand base $|\mathcal{B}|$ of a set of state transitions E. This n is also the number of possible heads of rules. Furthermore, n is also the maximum size of a rule, i.e. the number of literals in the body; a literal can appear at most one time in the body of a rule. For each head there are 3^n possible bodies: each literal can either be positive, negative or absent from the body. From these preliminaries we conclude that the size of a NLP $|P|$ learned by **LF1T** from E is at most $n \cdot 3^n$. But since a NLP P learned by **LF1T** only contains prime rules of E, $|P|$ cannot exceed $n \cdot 2^n$; in the worst case, P contains only rules of size n where all literals appear and there is only $n \cdot 2^n$ such rules. If P contains a rule with m literals $(m < n)$, this rule subsumes 2^{n-m} rules which cannot appear in P. Finally, least specialization also ensures that P does not contain any pair of complementary rules, so that the complexity is further divided by n; that is, $|P|$ is bounded by $O(\frac{n \cdot 2^n}{n}) = O(2^n)$.

When **LF1T** infers a rule R^I_A from a transition $(I, J) \in E$ where $A \notin J$, it has to compare it with all rules in P to extract the ones that need to be specialized. This operation has a complexity of $O(|P|) = O(2^n)$. Since $|b(R^I_A)| = n$, according to Definition 5 the least specialization of a rule $R \in P$ can at most generate n different rules. In the worst case all rules of P with $\overline{h(R^I_A)}$ as head subsume R^I_A. There are possibly $2^n/n$ such rules in P, so that **LF1T** generates at most 2^n rules for each R^I_A. For each $(I, J) \in E$, **LF1T** can infer at most n rules R^I_A. In the worst case, **LF1T** can generates $n \cdot 2^n$ rules that are compared with the 2^n rules of P. Thus, construction of an NLP which realizes E implies $n \cdot 2^n \cdot 2^n = n \cdot 4^n$

operations. The same proof applies to **LF1T** naïve (resp. ground) resolution, when **LF1T** infers a rule R_A^I from a transition $(I, J) \in E$ where $A \in J$. The complexity of learning an NLP from a complete set of state transitions with an Herbrand base of size n is $O(n \cdot 4^n)$. $\square$

References

1. Apt, K.R., Blair, H.A., Walker, A.: Towards a Theory of Declarative Knowledge. Foundations of Deductive Databases and Logic Programming, p. 89. Morgan Kaufmann, USA (1988)
2. Garcez, A., Zaverucha, G.: The connectionist inductive learning and logic programming system. Appl. Intel. **11**(1), 59–77 (1999)
3. Dubrova, E., Teslenko, M.: A sat-based algorithm for finding attractors in synchronous boolean networks. IEEE/ACM Trans. Computat. Biol. Bioinf. (TCBB) **8**(5), 1393–1399 (2011)
4. d'Avila Garcez, A.S., Broda, K., Gabbay, D.: Symbolic knowledge extraction from trained neural networks: a sound approach. Artif. Intel. **125**(2), 155–207 (2001). http://www.sciencedirect.com/science/article/pii/S0004370200000771
5. Gebser, M., Kaminski, R., Kaufmann, B., Schaub, T.: Answer Set Solving in Practice. Synthesis Lectures on Artificial Intelligence and Machine Learning. Morgan and Claypool Publishers, San Rafael (2012)
6. Inoue, K.: Logic programming for boolean networks. In: Proceedings of the Twenty-Second International Joint Conference on Artificial Intelligence-Volume, vol. 2, pp. 924–930. AAAI Press (2011)
7. Inoue, K.: DNF hypotheses in explanatory induction. In: Muggleton, S.H., Tamaddoni-Nezhad, A., Lisi, F.A. (eds.) ILP 2011. LNCS, vol. 7207, pp. 173–188. Springer, Heidelberg (2012)
8. Inoue, K., Ribeiro, T., Sakama, C.: Learning from interpretation transition. Mach. Learn. **94**(1), 51–79 (2014)
9. Inoue, K., Sakama, C.: Oscillating behavior of logic programs. In: Erdem, E., Lee, J., Lierler, Y., Pearce, D. (eds.) Correct Reasoning. LNCS, vol. 7265, pp. 345–362. Springer, Heidelberg (2012)
10. Kean, A., Tsiknis, G.: An incremental method for generating prime implicants/implicates. J. Symb. Comput. **9**(2), 185–206 (1990)
11. Michalski, R.S.: A theory and methodology of inductive learning. Artif. Intel. **20**(2), 111–161 (1983)
12. Mitchell, T.M.: Generalization as search. Artif. Intel. **18**(2), 203–226 (1982)
13. Muggleton, S., De Raedt, L.: Inductive logic programming: theory and methods. J. Logic Program. **19**, 629–679 (1994)
14. Muggleton, S., De Raedt, L., Poole, D., Bratko, I., Flach, P., Inoue, K., Srinivasan, A.: Ilp turns 20. Mach. Learn. **86**(1), 3–23 (2012)
15. Plotkin, G.D.: A note on inductive generalization. Mach. Intel. **5**(1), 153–163 (1970)
16. Ribeiro, T., Inoue, K., Sakama, C.: A BDD-based algorithm for learning from interpretation transition. In: Zaverucha, G., Santos Costa, V., Paes, A. (eds.) ILP 2013. LNCS, vol. 8812, pp. 47–63. Springer, Heidelberg (2014)
17. Tison, P.: Generalization of consensus theory and application to the minimization of boolean functions. IEEE Trans. Electron. Comput. **4**, 446–456 (1967)
18. Van Emden, M.H., Kowalski, R.A.: The semantics of predicate logic as a programming language. J. ACM (JACM) **23**(4), 733–742 (1976)

Statistical Relational Learning
for Handwriting Recognition

Arti Shivram[1]([✉]), Tushar Khot[2], Sriraam Natarajan[3], and Venu Govindaraju[1]

[1] University at Buffalo - SUNY, Buffalo, USA
ashivram@buffalo.edu
[2] University of Wisconsin-Madison, Madison, USA
[3] Indiana University, Bloomington, USA

Abstract. We introduce a novel application of handwriting recognition for Statistical Relational Learning. The proposed framework captures the intrinsic structure of handwriting by modeling fundamental character shape representations and their relationships using first-order logic. Our framework consists of three stages, (1) character extraction (2) feature generation and (3) class label prediction. In the character extraction stage, handwriting trajectory data is decoded into characters. Following this, character features (predicates) are defined across multiple levels - global, local and aggregated. Finally, a relational One-vs-All classifier is learned using relational functional gradient boosting (RFGB). We evaluate our approach on two datasets and demonstrate comparable accuracy to a well-established, meticulously engineered approach in the handwriting recognition paradigm.

1 Introduction

The prevalence of structural, noisy data in many domains has driven the recent development of *Statistical Relational AI* that combine statistical AI methods with relational or logical models. At the core of Statistical Relational AI is Statistical Relational Learning (SRL) [1], which addresses the challenge of applying statistical inference and learning approaches to problems which involve rich collections of objects varying in number and linked together in a complex, stochastic and relational world. The advantage of these models is that they can succinctly represent dependencies among the attributes of related objects, leading to a compact representation while exploiting the power of probability theory in modeling noise and uncertainty in the data.

Consequently, there is a plethora of real-world applications over the past few years. One of the top applications is *Natural Language Processing*, in particular information extraction [2–5] where SRL-based algorithms have been employed successfully. On the other side, biomedical tasks such as breast cancer [6] and drug activity prediction [7] have been addressed using powerful learning methods such as SAYU [8]. Recently, another learning method called Relational Functional Gradient Boosting (RFGB) [9] has been applied to heart attack prediction [10], cardiovascular risk prediction [11] and Alzheimer's disease prediction [12]. Finally, similar algorithms have also been used in vision tasks - see

© Springer International Publishing Switzerland 2015
J. Davis and J. Ramon (Eds.): ILP 2014, LNAI 9046, pp. 126–138, 2015.
DOI: 10.1007/978-3-319-23708-4_9

for instance the work of Antanas et al. [13,14]. Most of these approaches can be divided into two groups - ones where prior knowledge is used to guide the reasoning process [5,13,15] against the ones where the models are completely learned from data [7,10–12]. The latter approach is useful in situations where there is not significant prior knowledge to help learning a better model.

We consider this knowledge-free approach in an important task – *handwriting recognition*. In the field of handwriting recognition, digital input is captured as a temporal sequence of 2D points (x, y coordinates). The task of handwriting recognition is to map these sequence of points/pixels to either characters, words, sentences and so on. Thus, it can be addressed as a pattern recognition problem, and in most cases, a supervised sequential learning problem [16]. Despite the advances made in several handwriting systems [17–19], the key issue is that they are well-engineered solutions to the task at hand.

We, on the other hand, take a generalizable approach where we represent dependencies across multiple classes of objects by using a symbolic representation (predicate logic). This allows us to define properties (predicates) at the global character level as well as the local segment level in addition to aggregated properties (both global and local). The use of richer features allows us to reason between different classes. Specifically, we use RFGB along with a One-vs-All (OvA) strategy, where a classifier is learned for each class separately using all the other classes as negative examples and each class is discriminated from the others.

We make the following key contributions: first, we introduce the complex problem of unconstrained handwriting recognition to the ILP community. Second, we adapt a recently developed SRL algorithm for learning in this complex task. Third, we demonstrate the effectiveness of the proposed approach to recognition on real data sets by comparing it against a traditional well-engineered solution for this task.

The rest of the paper is organized as follows - after reviewing related work on handwriting recognition, we present our approach for solving this problem. Then we present experiments on two different data sets, before concluding by outlining problems for future research.

2 Background and Related Work

2.1 Handwriting Recognition

Current research in digital handwriting recognition may be classified into five broad approaches:

Template Based Models Template based models perform recognition by matching a given handwritten sequence with predefined prototypes, e.g., by matching substrokes to templates [20] or amount of deformation [21].

Motor Models Motor models [22], which are synonymous to an *analysis by synthesis* strategy, try to emulate the human motor movements involved in writing a particular pattern.

Structural or Syntactic Methods In syntactic methods [23–26] recognition is performed by enumerating the structural characteristics of the handwritten pattern. This is done by generating a rule-base describing the character shapes using a variable number of symbolic features.

Neural Network Architectures Neural network architectures [17,27] are often used to identify character segments in a word as a sliding window passes over the temporal sequence of points. Features extracted from points in the window frame are passed to the input layer of the network. Output node activations model the likelihood that the handwritten segment match the corresponding output node label. Most of the successful in-house handwriting recognition systems are based on neural networks and hence they are among the most popular methods for this task.

Stochastic Models Hidden Markov models (HMM) [18] and Conditional Random Fields (CRF) [19] are popular graphical models used for handwriting recognition. Given a sequence of features, the likelihood of it being generated is computed and the class with the highest likelihood value is returned.

2.2   RFGB

While successful, propositional approaches to digital handwriting recognition would treat all primitives such as lower-case characters, upper-case characters, numerals, punctuations, symbols etc. as being drawn independently from a single class of objects. In contrast, an SRL approach would not only allow us to treat each group of the above mentioned character types as a different class, but also provide a framework to *relate* how characters from one class affect the appearance or prevalence of characters from another class. In this way, a relational learning framework offers the potential to model richer dependencies probabilistically.

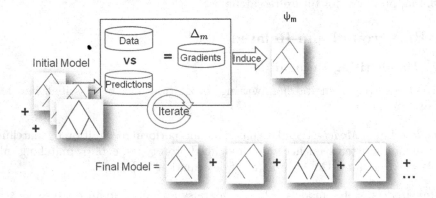

Fig. 1. Relational functional gradient boosting

Multiple SRL models have been proposed in the literature to model stochastic dependencies between the classes (relations/predicates) [28–31]. Parameter learning approaches rely on these dependencies being specified by an expert and learn the parameters from data [32,33]. But given the large number of handwritten patterns corresponding to a character, providing all the possible dependencies may not be possible. Hence, we need a structure learning approach for SRL models to learn the underlying dependencies along with the parameters. We use the recently successful structure learning approach, Relational Functional Gradient Boosting (RFGB) [9] that uses functional-gradient boosting to learn the structure and parameters of the conditional distributions. RFGB learns the conditional distribution of relational predicates given all the other predicates using Friedman's functional-gradient boosting [34] that uses a non-parametric representation for the conditional distributions and computes gradients over the functional space instead of the parametric space. RFGB maximizes the log-likelihood of the data given by the sum of the log-likelihoods of individual examples (ground atoms). The conditional distribution of an example is represented using the sigmoid over a regression function $\psi(x_i)$ where x_i is a ground example. To maximize the log-likelihood, gradient descent is performed over the functional space by calculating the gradients w.r.t. $\psi(x_i)$:

$$\Delta(x_i) = \frac{\partial \sum_i \log P(x_i; \mathbf{Pa(x_i)})}{\partial \psi(x_i; \mathbf{Pa(x_i)})} = I(x_i) - P(x_i; \mathbf{Pa(x_i)}) \qquad (1)$$

Given the regression values associated with individual relational examples, RFGB learns relational regression trees $(\hat{\psi}_j)$ to fit to these gradients. Similar to parametric descent, the final model is given by the sum of the individual gradients i.e. regression trees $(\sum_j \hat{\psi}_j)$. Similar to boosting approaches, RFGB learns a sequence of models where every subsequent model tries to correct the mistakes in the previous model. The RFGB approach is also shown in Fig. 1 and we refer to earlier work [9] for more details on the formalism.

3   Adapting RFGB to Handwriting Recognition

We present our three-stage approach to handwriting recognition in Fig. 2 - first word level raw data is preprocessed and the constituent characters are extracted; second global, local and aggregate features are generated for character patterns which are finally used in a one-vs-all learning framework. We now outline each of these stages in detail.

3.1   Word Level Preprocessing and Character Extraction

Handwriting capture via digital devices often introduces a small amount of noise causing trajectory distortions such as stray points, hooks, jagged edges etc. [35]. To correct for this, we first apply equidistant re-sampling of points along the stroke trajectory, followed by stroke smoothing. We perform equidistant re-sampling so that sampled pen-tip coordinates are equidistant in space rather

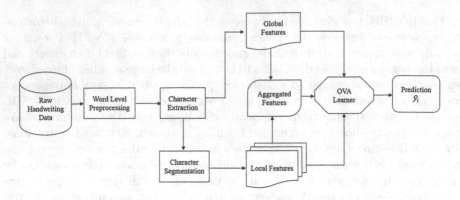

Fig. 2. Graphical representation of the pipeline

than time. We perform stroke smoothing to eliminate jagged edges along the
handwritten path, so as to create a smoothed uniform curve. We accomplish
this by applying a Gaussian filter independently to each coordinate point in the
sequence [35]:

$$x_i{}^{filtered} = \sum_{i=-3\sigma}^{3\sigma} w_i x_{t+i}^{orig}, \tag{2}$$

where

$$w_i = \frac{e^{-i^2/2\sigma^2}}{\sum_{j=-3\sigma}^{3\sigma} e^{-j^2/2\sigma^2}} \tag{3}$$

In addition to the above, unconstrained handwriting often suffers from varying slants introduced by different writers and across different points in a document. We carry out slant correction of word samples by shear transforming every
word according to it's slant angle [27]. Further, we size normalize each word such
that similar characters have approximately the same height. Finally, we extract
constituent character patterns from the data.

3.2   Generating Features

We now seek to identify a set of basic structural units - predicates - that each
capture positional (topological) and/or shape (geometric) characteristics of a
handwritten letter. Our goal is to demonstrate that features that model the general structure of the characters are powerful and are sufficiently predictive when
compared to the well-engineered solutions. These features are defined across
three levels called - global, local and aggregate features - here. Specifically, (1)
global features: correspond to properties attributable to the entire letter (character) (2) local features: correspond to properties attributable to segments of
the letter and (3) aggregate features: correspond to features derived from basic
global and local properties.

In order to define topological (position) and certain shape predicates we need to establish a spatial frame of reference. The boundaries of the letter shape, i.e., it's 'bounding box' forms this frame of reference. For instance, the left-most point of the letter (x_{min}) has the value zero and the right-most point of the letter (x_{max}) has value one. Similarly for the vertical dimension, y_{min} is zero and y_{max} is one. Thus, we get positional values that range between 0 and 1. These are discretized into three zones each for the vertical (top, center, bottom) and horizontal (left, middle, right) dimensions.

Inspired by techniques outlined in extant handwriting recognition research [24,36], we define the following feature predicates:

Global Features Global features help describe the letter as a whole and in a sense, represent the global approximation of the letter shape. To elaborate, we capture topological characteristics such as the start and end points of letter trajectories, handwritten shape characteristics such as the relative horizontal and vertical motion of the written letter profile and the aspect ratio of the letter shape [24,36]. We further define structural features such as the number of strokes (i.e., number of 'pen-ups'), the number of points within each stroke, the number of 'holes' (i.e., closed loops) and the relative positions of such holes.

Local Features For this level, each letter is broken down into segments, the properties of which form our local features. Each letter is segmented into primitives (or sub-strokes) on the basis of breakpoints that are defined by the sharpness in the curvature of the pen-tip trajectory. We locate these breakpoints by adapting the curvature metric proposed in [27]. At this point it must be highlighted that each letter need not have the same number of segments, leading to varying feature lengths across different character samples.

Subsequently, following [36] we measure local segmental properties like (1) 'arced'-ness, (2) 'straight'-ness, (3) whether the segment is a vertical line/curve, (4) is a horizontal line/curve, (5) is positively or negatively sloped, (6) the segment's horizontal convexity (how 'C/D-like' it is), (7) vertical convexity (how 'A/U-like' it is), (8) how 'O-like' a segment is and so on. In addition, a local positional feature - the relative location of the segment centroid - is also measured in order to capture the topological bias of the segment.

In our formulation all global and local properties provide real values in the range 0 to 1, which are then discretized based on histograms. For instance, the horizontal and vertical motions are binned into three categories 'Low', 'Med', 'High' and the aspect ratio into 'Thin', 'Square' and 'Wide'. The local segmental shape level properties are discretized into eight bins ranging from 'Zero' to 'Very Very High'. These discretizations are then used as arguments for the corresponding feature predicates.

Aggregated Features This feature type is defined by specifying certain combinations of the properties described earlier, a step that enables capturing finer-grained feature interactions. For example, an aggregation over a shape predicate (say, 'O-like') and a positional predicate (say, 'top-left' vs. 'bottom-right') allows us to distinguish between a 'q' vs. a 'b'. These aggregated predicates allow us to specify expert-defined relevant features to guide the learner.

Table 1 presents some of the predicates used in our learning experiments along with a brief explanation of these predicates.

3.3   One-vs-All Learner: RFGB

Once the appropriate predicates are designed, the next step is to learn a relational classifier (RFGB) for every letter in the vocabulary (the 26 $a - z$ letters). More precisely, the goal is to predict the label (the letter) given the predicates that we defined earlier in the image. We train the model for one letter by using the characters for that letter as positive examples and all the other characters as negative examples (One-vs-All). For inference, we use the probabilities returned by all the models. We then rank the letters based on their posterior distributions and return the character with the highest probability (MAP estimate).

The input to RFGB is the set of instances of the current letter (say a) which form the positive examples and the evidence (i.e., all the predicates defined earlier) associated with this letter. We learn 40 trees for each letter (resulting in a total of around 1040 trees). In every iteration of RFGB, the algorithm chooses

Table 1. A sample set of predicates used for handwriting recognition.

Type	Predicate	Explanation
Global (character level properties)	property_SX(C, X)/ property_SY(C, Y)	The start point of character C in the x-coordinate is X/The start point of character C in the y-coordinate is Y
	property_EX(C, X)/ property_EY(C, Y)	The end point of character C in the x-coordinate is X/The end point of character C in the y-coordinate is Y
	property_AR(C, A)	Aspect ratio of C is A
	property_numHoles(C, N)	The number of holes (closed loops) in C is N
Local (segment level properties)	segInChar(C, S)	Character C has segment S
	property_ARC(S, R)	The 'arced-ness' measure of S is R
	property_ST(S, T)	The 'straightness' measure of S is T
	property_C(S, E)/property_D(S, E)	The 'C-like'/'D-like' measure of S is E (convexity direction left/right)
	property_A(S, F)/property_U(S, F)	The 'A-like'/'U-like' measure of S is F (convexity direction down/up)
	property_PX(S, X)/property_PY(S, Y)	The centroid position of segment S in the x-coordinate is X/The centroid position of segment S in the y-coordinate is Y
Global aggregated	property_S_TL(C)	The start point of C is to the top - left
	property_E_BR(C)	The end point of C is to the bottom - right
Local aggregated	property_OL_LT(C)	C has a 'O-like' segment at the top - left
	property_U_R(C)	C has a 'U-like' segment to the right

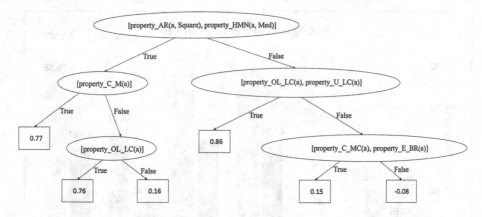

Fig. 3. The first tree learned in the RGFB forest for letter 'a'

a small sample from the other classes as negative examples. In our experiments, we used a pos:neg ratio of 1:2 and hence for every positive letter, RFGB randomly choses two other letters as negative examples for that iteration. Hence, in different iterations, RFGB can see different negative instances for the current positive instance.

More concretely, the OvA strategy employs 26 classifiers. Each of them discriminate the current label j from $j' \in class \setminus j$. We use a simple aggregation method called as *Maximum confidence strategy*. Each classifier c^k outputs the probability that the current letter is k and the output class is taken from the classifier that has the largest posterior probability $argmax_c\, p_c$. For more details on the OvA aggregation, please refer to [37].

Figure 3 shows a sample tree learned by RFGB for the letter 'a'. The regression tree returns a positive value (> 0.5 probability) if the character contains a C-like segment in the middle (property_C_M) or O-like segment to the left and center (property_OL_LC).

4   Experiments

In order to evaluate the performance of our framework, we undertake a comparative study vis-a-vis a baseline relational model as well as an established Neural Network-based approach from the handwriting recognition paradigm. In the following section, we first outline the datasets used for this comparison before presenting the results.

We utilized two publicly available handwriting datasets IBM_UB_1 [38] and UNIPEN [39]. IBM_UB_1 consists of a summary and query page for a set of authors. The summary is a full page of unconstrained cursive writing whereas the query page consists of approximately 25 handwritten words from the summary. We used only the query pages since they have word-level granularity.

We consider recognition of handwritten characters (specifically lower-case letters) as a first step towards recognition of written words. Since IBM_UB_1 is a very large dataset with more than 400,000 character samples, in the interest of

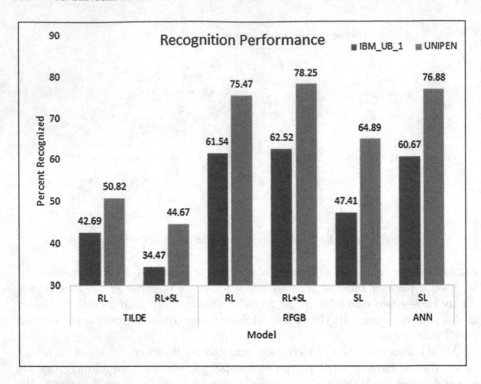

Fig. 4. Recognition performance across models. The blue and red bars denote the IBM and UNIPEN datasets. We present two SRL models (TILDE learner and RFGB) and a neural network learner on these tasks. RL denotes using only relational predicates, SL denotes the use of stroke-level predicates. We used RL and RL+SL for the SRL methods while we use only SL features for ANN (Color figure online).

computational tractability we use a smaller subset for our experimental setup. Using the train/test split provided in this dataset, we randomly pick 100 samples per class from the training subset and 50 samples per class from the test subset. Thus, we train on 2600 character samples (100 per class and 26 classes) and test on 1300 character samples maintaining a 2:1 ratio between train/test. We use three train/test folds for our evaluation.

UNIPEN, on the other hand, consists of handwritten isolated characters, printed and cursive words. For the current research, we use a subset of cursive words that contain character level segmentation written by 44 writers (details outlined in [19]). Since this subset does not contain a standardized train/test split, we randomize selection by author. We train on words written by one half of the writers (22 randomly selected writers) and test on words written by the other half. Similar to IBM_UB_1, we create three such folds for our evaluation.

The questions that we evaluate in our empirical study are:

1. How does RFGB compare to other relational approaches on these datasets?
2. How does RFGB compare to extant approaches in handwriting recognition?
3. How robust is the relational framework across different data sources?

Our basic experimental design involved two benchmarks: first, we compared RFGB against the baseline relational approach - TILDE tree and second, we compared RFGB against a mainstream recognition approach - feed-forward neural networks [40]. We conducted these comparisons using both relational (RL) and stroke-level (SL) features. The former refers to the predicates described earlier in this paper whereas the latter is composed of directional histogram features of the stroke trajectories [19,41].

It must be highlighted here that for the latter comparison, i.e., RFGB vs NN, given theoretical constraints of the two approaches, identical feature sets could not be used. For example, the relational features (RL) are of varying lengths for each character sample and hence do not have a one-on-one mapping to a propositional set-up. Directional histogram of stroke trajectories, on the other hand, are robust and highly effective features for character recognition and have also been used in conjunction with NNs [40,42], giving us a well established propositional benchmark to compare against. However, in order to ensure completeness, we evaluate the performance of the RFGB model when using stroke level features.

The baseline TILDE classifier represents a single tree learned without boosting in the RFGB framework [9]. On the other hand, the NN classifier is implemented using the SNNS toolkit [43]. We build a fully connected network consisting of a single hidden layer. The network has 62 input nodes *(representing the stroke level feature dimension)* and 26 output node activations *(representing the number of character classes)*. Following the convention of averaging the number of input and output nodes, our hidden layer is made up of 44 nodes. The error back-propagation algorithm is used to learn the weights of the network and training is halted when the mean squared error (MSE) over the training samples asymptotes *(∼1000/1500 epochs)*. Figure 4 summarizes the results of our experimental comparisons.

As can be observed from the figure, the boosted SRL model (RFGB) outperforms the single tree learner (TILDE) on both datasets with the different sets of features. Also, it can be seen that RFGB algorithm is on par or better than the ANN on both datasets when using only the relational predicates described earlier (*RL* in the figure). Additionally, when using the stroke level features as well, it improves upon the propositional benchmark on both datasets. Note that since in both tasks, we do not have any prior knowledge, we could not compare against other parameter learning methods inside SRL. Also we were not able to get any other structure learning algorithm for RDNs to learn anything useful in both data sets.

Hence, questions 1 and 3 can be answered affirmatively, in that RFGB performs favorably against TILDE in both the data sets. And w.r.t. question 2, the answer is a bit more conservative in that RFGB performs on par or slightly (not statistically significant) better than the ANN method. This gives us hope in that use of simple relational features without much engineering has good potential in learning robust models on this challenging task.

5 Conclusion

We introduced a novel but important task to the ILP and SRL communities. We showed how the task of handwriting recognition can be naturally modeled using SRL and applied a recently developed algorithm to this task. Our results showed that with minimal feature engineering, a powerful SRL method can perform on par with mainstream recognition algorithms. The goal of this work is not to show that RFGB is a powerful method but to demonstrate that relational learning is well suited for this task. As mentioned earlier, feature engineering can be improved and better discretizations can potentially improve the performance. More importantly, incorporating (a) hierarchies to identify sets of characters, and, (b) dictionaries to build word-level models, offer interesting future directions. We believe that relational learning allows for jointly learning word and character models and consider the current work as a first-step in this direction.

References

1. Getoor, L., Taskar, B.: Introduction to Statistical Relational Learning. MIT Press, Cambridge (2007)
2. Riedel, S., Chun, H.W., Takagi, T., Tsujii, J.: A Markov logic approach to biomolecular event extraction. In: Proceedings of the Workshop on Current Trends in Biomedical Natural Language Processing: Shared Task, Association for Computational Linguistics, pp. 41–49 (2009)
3. Poon, H., Vanderwende, L.: Joint inference for knowledge extraction from biomedical literature. In: ACL, Association for Computational Linguistics, pp. 813–821 (2010)
4. Yoshikawa, K., Riedel, S., Asahara, M., Matsumoto, Y.: Jointly identifying temporal relations with Markov logic. In: ACL, pp. 405–413 (2009)
5. Poon, H., Domingos, P.: Unsupervised semantic parsing. In: EMNLP, Association for Computational Linguistics, pp. 1–10 (2009)
6. Davis, J., Burnside, E.S., de Castro Dutra, I., Page, D., Ramakrishnan, R., Costa, V.S., Shavlik, J.W.: View learning for statistical relational learning: with an application to mammography. In: IJCAI, DTIC Document, pp. 677–683 (2005)
7. Davis, J., Ong, I.M., Struyf, J., Burnside, E.S., Page, D., Costa, V.S.: Change of representation for statistical relational learning. In: IJCAI, pp. 2719–2726 (2007)
8. Davis, J., Burnside, E., de Castro Dutra, I., Page, D.L., Santos Costa, V.: An integrated approach to learning Bayesian networks of rules. In: Gama, J., Camacho, R., Brazdil, P.B., Jorge, A.M., Torgo, L. (eds.) ECML 2005. LNCS (LNAI), vol. 3720, pp. 84–95. Springer, Heidelberg (2005)
9. Natarajan, S., Khot, T., Kersting, K., Gutmann, B., Shavlik, J.: Gradient-based boosting for statistical relational learning: the relational dependency network case. Mach. Learn. **86**(1), 25–56 (2012)
10. Weiss, J., Natarajan, S., Peissig, P., McCarty, C., Page, D.: Statistical relational learning to predict primary myocardial infarction from electronic health records. In: AI Magazine (2012)
11. Natarajan, S., Kersting, K., Ip, E., Jacobs, D., Carr, J.: Early prediction of coronary artery calcification levels using machine learning. In: Innovative Applications in AI (2013)

12. Natarajan, S., Saha, B., Joshi, S., Edwards, A., Khot, T., Davenport, E.M., Kersting, K., Whitlow, C.T., Maldjian, J.A.: Relational learning helps in three-way classification of Alzheimer patients from structural magnetic resonance images of the brain. Int. J. Mach. Learn. Cybern. **5**(5), 659–669 (2013)
13. Antanas, L., van Otterlo, M., Mogrovejo, O., Antonio, J., Tuytelaars, T., De Raedt, L.: A relational distance-based framework for hierarchical image understanding. In: Proceedings of the 1st International Conference on Pattern Recognition Applications and Methods, vol. 2, pp. 206–218 (2012)
14. Antanas, L., Hoffmann, M., Frasconi, P., Tuytelaars, T., De Raedt, L.: A relational kernel-based approach to scene classification. In: 2013 IEEE Workshop on Applications of Computer Vision (WACV), pp. 133–139, IEEE (2013)
15. Riedel, S., McClosky, D., Surdeanu, M., McCallum, A., Manning, C.D.: Model combination for event extraction in bionlp 2011. In: Proceedings of the BioNLP Shared Task 2011 Workshop, Association for Computational Linguistics, pp. 51–55 (2011)
16. Dietterich, T.G.: Machine learning for sequential data: a review. In: Caelli, T.M., Amin, A., Duin, R.P.W., Kamel, M.S., de Ridder, D. (eds.) SPR 2002 and SSPR 2002. LNCS, vol. 2396, pp. 15–30. Springer, Heidelberg (2002)
17. Graves, A., Liwicki, M., Bunke, H., Schmidhuber, J., Fernández, S.: Unconstrained on-line handwriting recognition with recurrent neural networks. In: Platt, J.C., Koller, D., Singer, Y., Roweis, S.T. (eds.) Advances in Neural Information Processing Systems 20, pp. 577–584. MIT Press, Cambridge (2007)
18. Liwicki, M., Bunke, H., et al.: HMM-based on-line recognition of handwritten whiteboard notes. In: Tenth International Workshop on Frontiers in Handwriting Recognition (2006)
19. Shivram, A., Zhu, B., Setlur, S., Nakagawa, M., Govindaraju, V.: Segmentation based online word recognition: a conditional random field driven beam search strategy. In: 2013 12th International Conference on Document Analysis and Recognition (ICDAR), pp. 852–856, IEEE (2013)
20. Guberman, S.A., Lossev, I., Pashintsev, A.V.: Method and apparatus for recognizing cursive writing from sequential input information, 17 May 1994. US Patent 5,313,527
21. Li, X., Yeung, D.Y.: On-line handwritten alphanumeric character recognition using dominant points in strokes. Pattern Recogn. **30**(1), 31–44 (1997)
22. Plamondon, R., Maarse, F.J.: An evaluation of motor models of handwriting. IEEE Trans. Syst. Man Cybern. **19**(5), 1060–1072 (1989)
23. Parizeau, M., Plamondon, R.: A fuzzy-syntactic approach to allograph modeling for cursive script recognition. IEEE Trans. Pattern Anal. Mach. Intell. **17**(7), 702–712 (1995)
24. Malaviya, A., Peters, L.: Fuzzy handwriting description language: Fohdel. Pattern Recogn. **33**(1), 119–131 (2000)
25. Ziino, D., Amin, A., Sammut, C.: Recognition of hand printed Latin characters using machine learning. In: Proceedings of the Third International Conference on Document Analysis and Recognition 1995, vol. 2, pp. 1098–1102, IEEE (1995)
26. Amin, A., Sammut, C., Sum, K.: Learning to recognize hand-printed Chinese characters using inductive logic programming. Int. J. Pattern Recogn. Artif. Intell. **10**(07), 829–847 (1996)
27. Manke, S., Finke, M., Waibel, A.: NPen++: a writer independent, large vocabulary on-line cursive handwriting recognition system. In: Proceedings of the Third International Conference on Document Analysis and Recognition 1995, vol. 1, pp. 403–408, August 1995

28. Neville, J., Jensen, D.: Relational dependency networks. In: Getoor, L., Taskar, B. (eds.) Introduction to Statistical Relational Learning. MIT Press, Cambridge (2007)
29. Domingos, P., Lowd, D.: Markov logic: an interface layer for artificial intelligence. Synth. Lect. Artif. Intell. Mach. Learn. 3(1), 1–155 (2009)
30. Kersting, K., De Raedt, L.: Bayesian logic programming: theory and tool. In: Getoor, L., Taskar, B. (eds.) Statistical Relational Learning, p. 291. MIT Press, Cambridge (2007)
31. Friedman, N., Getoor, L., Koller, D., Pfeffer, A.: Learning probabilistic relational models. In: IJCAI, vol. 99, pp. 1300–1309 (1999)
32. Singla, P., Domingos, P.: Discriminative training of Markov logic networks. In: AAAI, vol. 5, pp. 868–873 (2005)
33. Lowd, D., Domingos, P.: Recursive random fields. In: IJCAI, pp. 950–955 (2007)
34. Friedman, J.H.: Greedy function approximation: a gradient boosting machine. Ann. Stat. 29(5), 1189–1232 (2001)
35. Connell, S.D.: Online handwriting recognition using multiple pattern class models. Ph.D. thesis (2000)
36. Malaviya, A., Peters, L.: Fuzzy feature description of handwriting patterns. Pattern Recogn. 30(10), 1591–1604 (1997)
37. Galar, M., Fernández, A., Barrenechea, E., Bustince, H., Herrera, F.: An overview of ensemble methods for binary classifiers in multi-class problems: experimental study on one-vs-one and one-vs-all schemes. Pattern Recogn. 44(8), 1761–1776 (2011)
38. Shivram, A., Ramaiah, C., Setlur, S., Govindaraju, V.: IBM_UB_1: a dual mode unconstrained English handwriting dataset. In: Proceedings of the Twelfth International Conference on Document Analysis and Recognition 2013, pp. 13–17, (2013)
39. Guyon, I., Schomaker, L., Plamondon, R., Liberman, M., Janet, S.: Unipen project of on-line data exchange and recognizer benchmarks. In: Proceedings of the 12th IAPR International. Conference on Pattern Recognition 1994. Vol. 2-Conference B: Computer Vision &; Image Processing, vol. 2, pp. 29–33, IEEE (1994)
40. Liu, C.L., Sako, H., Fujisawa, H.: Performance evaluation of pattern classifiers for handwritten character recognition. Int. J. Doc. Anal. Recogn. 4(3), 191–204 (2002)
41. Zhu, B., Shivram, A., Setlur, S., Govindaraju, V., Nakagawa, M.: Online handwritten cursive word recognition using segmentation-free MRF in combination with P2DBMN-MQDF. In: 2013 12th International Conference on Document Analysis and Recognition (ICDAR), pp. 349–353, IEEE (2013)
42. Cao, J., Shridhar, M., Kimura, F., Ahmadi, M.: Statistical and neural classification of handwritten numerals: a comparative study. In: 11th IAPR International Conference on Pattern Recognition 1992, vol. II. Conference B: Pattern Recognition Methodology and Systems, Proceedings, pp. 643–646, IEEE (1992)
43. SNNS toolkit. http://www.ra.cs.uni-tuebingen.de/SNNS/

The Most Probable Explanation for Probabilistic Logic Programs with Annotated Disjunctions

Dimitar Shterionov[(✉)], Joris Renkens, Jonas Vlasselaer, Angelika Kimmig, Wannes Meert, and Gerda Janssens

KULeuven, Leuven, Belgium
Dimitar.Shterionov@cs.kuleuven.be

Abstract. Probabilistic logic languages, such as ProbLog and CP-logic, are probabilistic generalizations of logic programming that allow one to model probability distributions over complex, structured domains. Their key probabilistic constructs are probabilistic facts and annotated disjunctions to represent binary and mutli-valued random variables, respectively. ProbLog allows the use of annotated disjunctions by translating them into probabilistic facts and rules. This encoding is tailored towards the task of computing the marginal probability of a query given evidence (MARG), but is not correct for the task of finding the most probable explanation (MPE) with important applications e.g., diagnostics and scheduling.

In this work, we propose a new encoding of annotated disjunctions which allows correct MARG and MPE. We explore from both theoretical and experimental perspective the trade-off between the encoding suitable only for MARG inference and the newly proposed (general) approach.

Keywords: Probabilistic logic programming · Statistical relational learning · Most Probable Explanation · Logic programs with annotated disjunctions · ProbLog

1 Introduction

The field of Probabilistic Logic Programming (PLP) combines probabilistic reasoning and logic to deal with complex relational domains. Most PLP techniques extend logic programming languages (such as Prolog) with probabilities. This has resulted in different probabilistic logic languages and systems, such as ProbLog [1], PRISM [2], LPADs [3], PHA [4] and others, which mostly focus on computing marginal probabilities of individual random variables or queries.

In contrast, the field of Statistical Relation Learning (SRL) often focuses on the extension of graphical models with logical and relational representations. Here, the most common inference tasks are computing the marginal probability of a set of random variables given evidence (MARG) and finding the most probable explanation given evidence (MPE).

The MPE task did not yet receive much attention in PLP, although it has important applications in diagnostics and scheduling. For a diagnostics example,

© Springer International Publishing Switzerland 2015
J. Davis and J. Ramon (Eds.): ILP 2014, LNAI 9046, pp. 139–153, 2015.
DOI: 10.1007/978-3-319-23708-4_10

it is equivalent to finding the health state of a set of machines or components given a set of measurements. For a scheduling problem, one is interested in an optimal way of distributing resources or tasks. MPE is a special case of the computationally harder MAP (maximum a posteriori) task. The MPE is different from the task of finding the most probable proof in PLP, also called Viterbi proof [4–7], where one is interested in the proof to a query with highest probability.

ProbLog [1,8] is a probabilistic logic and learning framework – a language and an engine. The engine handles various inference and learning tasks, including MPE inference [9], and as such tries to bridge the gap between PLP and SRL. Initially focused on probabilistic facts, or binary random variables, the ProbLog language now also supports Annotated Disjunctions (ADs) [3]. These specify exclusive choices between alternatives given some condition and thus provide an intuitive construct to represent multi-valued random variables. Internally, ProbLog encodes ADs by means of probabilistic facts and rules. However, given the semantics of ADs, this encoding is not correct for the MPE task.

In this paper, we propose an encoding of Annotated Disjunctions that is correct for both MARG and MPE inference. It uses ProbLog constraints, that is, first order logic sentences which restrict the models of ProbLog programs [10]. We compare the two encodings both theoretically and experimentally. We also incorporate an MPE inference algorithm based on weighted model counting into ProbLog.

The paper is organized as follows: In Sect. 2 we give background on the language ProbLog, introduce the MPE task and show why the encoding of ADs as ProbLog programs is not correct for the MPE task. Section 3 introduces our approach to encode ADs as weighted CNFs and proves its correctness. Section 4 summarizes experimental results. We conclude in Sect. 5.

2   Background

2.1   The ProbLog Language

The ProbLog language [1] is a probabilistic extension of Prolog. It uses probabilistic facts, that is, facts whose truth values are determined probabilistically to model basic uncertain data, and Prolog rules to define deterministic consequences of the probabilistic facts. A probabilistic fact has the form $p_i : : f_i$ and states that the fact f_i is true with probability $true(f_i) = p_i$, and false otherwise (with $false(f_i) = (1 - p_i)$). ProbLog uses the *distribution semantics* [11] to define a probability distribution over the least models (or *possible worlds*) of the ProbLog program. We write a possible world ω as a tuple (ω^+, ω^-), where the set ω^+ contains the probabilistic facts that are true and ω^- those that are false, thus, omitting the (uniquely determined) truth values for non-probabilistic atoms. Treating probabilistic facts as independent binary random variables, we obtain the probability of a possible world as:

$$P(\omega) = \prod_{f_i \in \omega^+} true(f_i) \prod_{f_i \in \omega^-} false(f_i) \qquad (1)$$

Example 1. The following ProbLog program models a game with three bags with colored balls. One ball is picked from each bag, and the game is won if at least two balls are red. The probabilistic fact `0.6::red(b1)` states that the probability of selecting a red ball from bag `b1` is 0.6. It implies that selecting a ball with another color (e.g. blue or green) has probability 0.4. The table outlines the set of its possible worlds (r_i abbreviates `red(bi)`) and their probabilities:

	poss. world	r_1	r_2	r_3	win	$P(r_1)$	$P(r_2)$	$P(r_3)$	$P(\omega_i)$
`0.6::red(b1).`	ω_1	T	T	T	T	0.6	0.2	0.7	0.084
`0.2::red(b2).`	ω_2	T	T	F	T	0.6	0.2	0.3	0.036
`0.7::red(b3).`	ω_3	T	F	T	T	0.6	0.8	0.7	0.336
`win:- red(b1), red(b2).`	ω_4	T	F	F	F	0.6	0.8	0.3	0.144
`win:- red(b2), red(b3).`	ω_5	F	T	T	T	0.4	0.2	0.7	0.056
`win:- red(b1), red(b3).`	ω_6	F	T	F	F	0.4	0.2	0.3	0.024
	ω_7	F	F	T	F	0.4	0.8	0.7	0.224
	ω_8	F	F	F	F	0.4	0.8	0.3	0.096

2.2 ProbLog Programs with Annotated Disjunctions

Annotated Disjunctions. We recall annotated disjunctions as used in LPADs [3] and CP-Logic [12]. An annotated disjunction (AD) a is a multi-headed rule of the form $p_1 : :h_1; \ldots; p_n : :h_n \leftarrow b_1, \ldots, b_m.$, where p_i is the probability of the head atom h_i, the p_i sum to at most one, and the conjunction of the literals $b_1, \ldots, b_m$ forms the body of the AD. We denote with $head(a)$ the set of head atoms of the AD a and with $body(a)$ the set of body literals of a. If the body is true, the AD probabilistically *causes* one of the head atoms to become true, otherwise the AD does not have an effect. Thus, if the same atom appears in the heads of multiple ADs, they correspond to different causes. If $\sum_{i=1}^{n} p_i < 1$ there is a probability $(= 1 - \sum_{i=1}^{n} p_i)$ that none of the head atoms is caused to be true. As this can be made explicit by adding an extra *none* head atom, we assume that probabilities sum to one.

We now summarize the process semantics of annotated disjunctions as defined in CP-logic; we refer to [12] for full technical details and a proof of the equivalence of this semantics with the instance based semantics of LPADs [3,12].

Definition 1 (Probability Tree). *Let $A = \{a_1, \ldots, a_k\}$ be a set of ground annotated disjunctions over atoms L_A. A probability tree $T(A)$ is a tree where every node n is labeled with an interpretation $I(n)$ assigning truth values to a subset of L_A and a probability $P(n)$, constructed as follows:*

- *The root node $\perp$ has probability $P(\perp) = 1.0$ and interpretation $I(\perp) = \{\}$.*
- *Each inner node n is associated with an AD a_i such that*
  - *no ancestor of n is associated with a_i,*
  - *all positive literals in $body(a_i)$ are true in $I(n)$,*
  - *for each negative literal $\neg l$ in $body(a_i)$, the positive literal l cannot be made true starting from $I(n)$.*

  and has one child node for each atom $h_j \in head(a_i)$. The jth child has interpretation $I(n) \cup \{h_j\}$ and probability $P(n) \cdot true(h_j)$.

– No leaf can be associated with an AD following the rule above.

The path from the root to a leaf n is called a selection σ_n with probability $P(\sigma_n) = P(n)$. We say that each selection defines an interpretation and denote with $I(\sigma_n)$ the interpretation defined by σ_n $(I(\sigma_n) = I(n))$.

All probability trees for a given set of ADs define the same distribution over selections [12]; from here on, we refer to an arbitrary probability tree when mentioning "the probability tree of A".

Example 2. We consider a variant of the game in Example 1 with a single bag containing red, green and blue balls, as expressed by AD a_1, where the player randomly decides to pick a ball or not, as encoded by AD a_2:

a_1: 0.6::red(b1); 0.3::green(b1); 0.1::blue(b1) <- pick(b1).

a_2: 0.6::pick(b1); 0.4::no_pick(b1)<- true.

A probability tree associated with these ADs and its selections are:

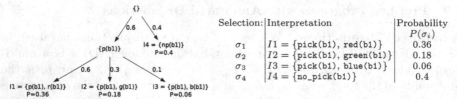

Selection:	Interpretation	Probability $P(\sigma_i)$
σ_1	$I1 = \{\text{pick(b1)}, \text{red(b1)}\}$	0.36
σ_2	$I2 = \{\text{pick(b1)}, \text{green(b1)}\}$	0.18
σ_3	$I3 = \{\text{pick(b1)}, \text{blue(b1)}\}$	0.06
σ_4	$I4 = \{\text{no\_pick(b1)}\}$	0.4

Here, all selections define different interpretations.

ProbLog Encoding of Annotated Disjunctions. To support MARG inference for ProbLog programs with annotated disjunctions, an AD is translated to a set of probabilistic facts with suitably adapted probabilities and a Prolog rule for each of its head atoms, with bodies which are mutually exclusive. We refer to [13, Chapter3] for the details and illustrate the principle with an example.

Example 3. The ProbLog encoding for Example 2:

```
0.6::pf(1, 1).      0.75::pf(1, 2).              0.6::pf(2, 1).
red(b1):- pick(b1), pf(1, 1).                    pick(b1):- pf(2, 1).
green(b1):- pick(b1), pf(1, 2), \+ pf(1, 1).     no_pick(b1):-\+ pf(2, 1).
blue(b1):- pick(b1), \+ pf(1, 2), \+ pf(1, 1).
```

2.3   Most Probable Explanations for ProbLog Programs

The inference task that has received most attention by the PLP community is computing the marginal probability of a ground query atom q given evidence $E = e$ on a subset of the other atoms, $\text{MARG}(q \mid E = e) = P(q \mid E = e)$. In the absence of evidence, this is also known as the success probability of a query.

In statistical relational learning (SRL) [14] and probabilistic graphical models [15] one of the key tasks is to find the most likely state of the world where a set of observations (the evidence) holds, also called MPE inference. Formally, the Most Probable Explanation (MPE) task, is defined as follows.

Definition 2 (Most Probable Explanation). *Given a probability distribution $P(V)$ over a set of discrete random variables V and a truth value assignment e to a subset of random variables $E \subseteq V$ (the evidence), the task of finding the most probable explanation (MPE) is to determine a truth value assignment u to the remaining random variables $U = V \setminus E$ with maximal probability, that is, $\mathrm{MPE}(E) = \arg\max_u P(U = u \mid E = e)$.*

A solution of the MPE task is also called an MPE state. Solution techniques developed for MPE include methods based on knowledge compilation [16], integer linear programming [17], and weighted MAX-SAT [18].

For a **ProbLog program without ADs**, V is the set of ground atoms, $E = e$ fixes truth values for a subset of those, and the task is to find the most likely assignment to all other atoms, that is, the possible world with the highest probability in which the evidence holds. As in [9], we assume ProbLog programs with finite groundings.

Example 4. For Example 1 given the evidence $\texttt{red(b2)} = true$ the worlds where the evidence holds are ω_1, ω_2, ω_5 and ω_6. The one with the highest probability is ω_1. The MPE state is thus $\{\texttt{red(b1)} = true, \texttt{red(b2)} = true, \texttt{red(b3)} = true\}$.

For a set of **annotated disjunctions** $A = \{a_1, .., a_k\}$, the most probable explanation is the truth value assignment according to the interpretation defined by the selection in $T(A)$ with highest probability amongst the ones for which the evidence holds. Let $\mathcal{S}^{E=e} = \{\sigma \mid I(\sigma) \models E = e\}$ be the set of selections in which the evidence holds and $\hat{\sigma} = argmax_{\sigma \in \mathcal{S}^{E=e}} P(\sigma)$, then $MPE_A(E) = I(\hat{\sigma})$.

Example 5. For the set of ADs in Example 2, the evidence $\texttt{blue(b1)} = false$ makes σ_3 invalid. The MPE state given the evidence thus is $\sigma_4 = \texttt{no}_p\texttt{ick(b1)}$.

The ProbLog encoding of these ADs (cf. Example 3) has the following possible worlds ($\texttt{p(b1)}$ is short for $\texttt{pick(b1)}$ and $\texttt{np(b1)}$ for $\texttt{no\_pick(b1)}$):

poss. world	pf(1, 1)	pf(1, 2)	pf(2, 1)	red(b1)	green(b1)	blue(b1)	p(b1)	np(b1)	$P(\omega_i)$
ω_1	T	T	T	T	F	F	T	F	0.27
ω_2	T	T	F	F	F	F	F	T	0.18
ω_3	T	F	T	T	F	F	T	F	0.09
ω_4	T	F	F	F	F	F	F	T	0.06
ω_5	F	T	T	F	T	F	T	F	0.18
ω_6	F	T	F	F	F	F	F	T	0.12
ω_7	F	F	T	F	F	T	T	F	0.06
ω_8	F	F	F	F	F	F	F	T	0.04

The evidence $\texttt{blue(b1)} = false$ holds in all worlds except ω_7, and the MPE state of the ProbLog program is thus $\{\texttt{pf}(1, 1) = true, \texttt{pf}(1, 2) = true, \texttt{pf}(2, 1) = true\}$. For this MPE state $\texttt{no}_p\texttt{ick(b1)}$ is false and $\texttt{pick(b1)}$ is true, which is different from the MPE state of the underlying set of ADs.

Example 5 illustrates that using the ProbLog encoding of annotated disjunctions can result in incorrect MPE states. The reason is that several possible worlds of the ProbLog encoding may correspond to the same selection in the probability tree. This is not a problem when computing marginal probabilities, as the probabilities of all possible worlds where the query is true are summed together.

In [9] MPE is mentioned as special case of the more general maximum a posteriori (MAP) inference. MAP is the task of finding the most likely values for a set of query atoms given *partial* evidence. The query atoms are a subset of all atoms of the ProbLog program for which evidence is not given. Solving then the MAP task requires to marginalize over the atoms which are neither given as queries nor as evidence. That is, we have to apply maximization over the query atoms given the evidence atoms and summation over the rest. MAP inference can be used in order to find the MPE state of a ProbLog program with ADs. For the ProbLog encoding of such a program we can give as queries all the atoms of that program excluding the probabilistic facts generated by the encoding. Then solving the MAP task will give the truth value assignments of these atoms which is in practice the MPE state of the initial program. The MAP inference is computationally expensive, while MPE can be computed efficiently [16,19]. The new encoding we introduce in Sect. 3.1 allows to solve the MPE task on ProbLog programs with ADs both correctly and efficiently by enforcing a one-to-one correspondence between selections and possible worlds.

Relation to Most Probable Proof. In PLP the term *most probable explanation* typically is used interchangeably with *most probable proof*, also called *Viterbi proof* [4–7]. A proof (or explanation) ω' for a query is a partial truth value assignment (or partial possible world) such that for all full assignments extending the proof, the query holds. Finding a most likely proof (the VIT task) is different from MPE in that it does not aim at finding the state of all unobserved variables, but an assignment to a small set of variables sufficient to explain a query. More formally, given a query q, we have $VIT(q) = \arg\max_{\omega' \in E(q)} P(\omega')$ with $E(q)$ the set of all explanations or proofs of q.

For certain types of models, finding the Viterbi proof for query q corresponds to solving MPE with $q = true$ as evidence. This is true for instance in programs modeling Hidden Markov Models, where both VIT and MPE have to make one choice per time point, but does not hold in general:

Example 6. In the program below, query win has two proofs: the first uses facts red and green and has probability $0.4 \cdot 0.9 = 0.36$, the second uses facts blue and yellow and has probability $0.5 \cdot 0.6 = 0.3$. The Viterbi proof for win is the first one. The MPE state for evidence win $= true$, however, is $\{\neg\text{red}, \text{green}, \text{blue}, \text{yellow}\}$. This state does not extend the Viterbi proof.

```
0.4::red.    0.9::green.       win :- red, green.
0.5::blue.   0.6::yellow.      win :- blue, yellow.
```

Relation to Most Probable Explanation for Bayesian Networks. To encode a Bayesian network using Annotated Disjunctions, each row in each conditional probability table (CPT) is encoded as a deterministic rule to capture the value assignments and a probabilistic fact to capture the probability. There is only one CPT per variable (the child node) and the rows in a CPT express mutually exclusive value assignments to a set of variables (the *parents* of the node). For each value assignment to the parents we introduce one AD with head atoms encoding the different value assignments to the child node and body associated to the value of the parents. The probability of each head atom is the probability given in the CPT for the value assignment of the parents. It has been shown that this encoding with ADs expresses the same probability distribution as the Bayesian network [20].

It can be shown that, given Definition 2, the MPE state of the Bayesian network is equivalent to the MPE state of a set of ADs that encode the Bayesian network. The probability tree inferred by a set of ADs that encode a Bayesian network has as property that each leaf has a unique interpretation. Therefore, the interpretation with the highest probability and the selection with the highest probability are equivalent. An intuitive proof can be constructed as follows: For any given CPT for a variable a, at each point in constructing the probability tree, if all parent variables of a certain CPT are present in the current partial interpretation, exactly one rule with a as head has a condition that is true. For each value of a a subtree is instantiated. None of the other rules can have a true condition in any of the subtrees due to the mutually exclusivity. This guarantees that each subtree has a different value assigned to a and no two interpretations in different subtrees can be identical.

3   Weighted CNF Encoding for Annotated Disjunctions

ProbLog inference is based on a transformation of the ProbLog program to a weighted Boolean formula in CNF, on which weighted model counting (WMC) is used to compute probabilities [8]. We now introduce a new encoding of annotated disjunctions in line with this transformation and consistent for MPE.

3.1   Encoding

The encoding of annotated disjunctions we propose (which we refer to as the *WMC* encoding to differentiate it from the *ProbLog* encoding discussed in Sect. 2.2) has two parts: (i) a logic program with weighted facts that is transformed to CNF as in ProbLog; (ii) a set of constraints that is directly added to the weighted CNF. This is a special case of ProbLog with constraints (cProbLog) [10], adapted directly to the specific constraints needed here. ProbLog employs weighted model counting (WMC) techniques for MARG and MPE inference on the weighted CNF.

Definition 3 (WMC encoding for ADs). *The WMC encoding of a ground AD* $p_1 ::h_1;..;p_n ::h_n \leftarrow b_1,..,b_m$. *with unique identifier* $\mathtt{a_j}$ *consists of:*

- *for each* h_i *with* $1 \leq i \leq n$ *a surrogate probabilistic fact* $\mathtt{spf(a_j,h_i,i)}$ *with* $true(\mathtt{spf(a_j,h_i,i)}) = p_i$ *and* $false(\mathtt{spf(a_j,h_i,i)}) = 1.0$ *and a clause*

$$h_i : -b_1,..,b_m, \mathtt{spf(a_j,h_i,i)}.$$

- *(the conjunction of) the following two constraints:*

$$\bigwedge_{i=1}^{n-1} \bigwedge_{l=i+1}^{n} (\neg \mathtt{spf(a_j,h_i,i)} \vee \neg \mathtt{spf(a_j,h_l,1)}), \qquad \bigwedge_{k=1}^{m} b_k \leftrightarrow \bigvee_{i=1}^{n} \mathtt{spf(a_j,h_i,i)}$$

Intuitively, surrogate probabilistic facts make the choices in ADs explicit, and constraints ensure that this does not introduce undesired combinations of values. When one head atom h_i is selected for an AD, the other head atoms are ignored. That is, they do not influence the probability of any selections with h_i true. That is why the false probability of a surrogate probabilistic fact is set to 1.0. The first constraint states that for each AD, at most one surrogate probabilistic fact can be true in any possible world, and thus only one head atom can be made true by the AD. The second constraint states that a choice is made if and only if the body of the AD is true. We write a surrogate probabilistic fact $\mathtt{spf}(\mathtt{a_j, h_i, i})$ with $true(\mathtt{spf}(\mathtt{a_j, h_i, i})) = p$ as $(p, 1.0) :: \mathtt{spf}(\mathtt{a_j, h_i, i})$.

Example 7. The WMC encoding of the two ADs of Example 2 consists of the program part:

```
(0.6, 1.0)::spf(1, red(b1), 1).           blue(b1):- pick(b1), spf(1, blue(b1), 3).
(0.3, 1.0)::spf(1, green(b1), 2).         (0.6, 1.0)::spf(2, pick(b1), 1).
(0.1, 1.0)::spf(1, blue(b1), 3).          (0.4, 1.0)::spf(2, no_pick(b1), 2).
red(b1):- pick(b1), spf(1, red(b1), 1).   pick(b1):- spf(2, pick(b1), 1).
green(b1):- pick(b1), spf(1, green(b1), 2).  no_pick(b1):- spf(2, no_pick(b1), 2).
```

and the four constraints where in the last one we omit the equivalence with *true*:

$(\neg\mathtt{spf}(1, \mathtt{red(b1)}, 1) \vee \neg\mathtt{spf}(1, \mathtt{green(b1)}, 2)) \wedge (\neg\mathtt{spf}(1, \mathtt{red(b1)}, 1) \vee \neg\mathtt{spf}(1, \mathtt{blue(b1)}, 3)) \wedge$

$\quad (\neg\mathtt{spf}(1, \mathtt{green(b1)}, 2) \vee \neg\mathtt{spf}(1, \mathtt{blue(b1)}, 3))$

$\mathtt{pick(b1)} \leftrightarrow (\mathtt{spf}(1, \mathtt{red(b1)}, 1) \vee \mathtt{spf}(1, \mathtt{green(b1)}, 2) \vee \mathtt{spf}(1, \mathtt{blue(b1)}, 3))$

$\neg\mathtt{spf}(2, \mathtt{pick(b1)}, 1) \vee \neg\mathtt{spf}(2, \mathtt{no\_pick(b1)}, 2)$

$\mathtt{spf}(2, \mathtt{pick(b1)}, 1) \vee \mathtt{spf}(2, \mathtt{no\_pick(b1)}, 2)$

The possible worlds of the program part in which the constraints hold are (r, g, b, p, np abbreviate `red(b1)`, `green(b1)`, `blue(b1)`, `pick(b1)`, `no_pick(b1)`):

poss. world	spf(1,r,1)	spf(1,g,2)	spf(1,b,3)	spf(2,p,1)	spf(2,np,2)	r	g	b	p	np	$P(\omega_i)$
ω_1	T	F	F	T	F	T	F	F	T	F	0.36
ω_2	F	T	F	T	F	F	T	F	T	F	0.18
ω_3	F	F	T	T	F	F	F	T	T	F	0.06
ω_4	F	F	F	F	T	F	F	F	F	T	0.40

3.2   Correctness

Theorem 1. *For a set $A = \{a_1, \ldots, a_k\}$ of ground annotated disjunctions, there is a bijection from the set $\mathcal{M}$ of models of the weighted CNF for A to the set $\mathcal{S}$ of selections in a probability tree T for A.*

Proof. *Let L_A be the set of atoms in A, and L_F the set of surrogate facts in the weighted CNF encoding of A. For every truth value assignment l_F to L_F, there is exactly one truth value assignment l_A to L_A such that $l_F \cup l_A$ is a model of the program part of the encoding [9]. The first constraint filters out all assignments l_F that assign true to more than one surrogate fact for the same ground AD, and the second filters out those that assign true to any surrogate* *fact for an AD whose body is false in $l_F \cup l_A$. Each remaining assignment $l_F \cup l_A$ is in one-to-one correspondence with a selection in S. That is, each such an assignment (i.e. a model) corresponds to exactly one path from the root to a leaf of the tree T (the assignment l_A is a model of the node's interpretation) and there is one model for each path.* ∎

Theorem 2. *Given a set $A = \{a_1, \ldots, a_k\}$ of ground annotated disjunctions, the set $\mathcal{M}$ of models of the weighted CNF for A, and the set S of selections in a probability tree T for A, the weight of a model $M \in \mathcal{M}$ equals the probability of the corresponding selection $S \in S$.*

Proof. *Follows directly from the fact that the model and the selection follow the same path through the tree and the definition of the weight function on the CNF. At each node, the probability of the selection so far is multiplied with the probability p_i of the chosen head atom, and the weight of the model with the weight of the AD's surrogate facts, $p_i \cdot 1.0 \cdot \ldots \cdot 1.0 = p_i$.* ∎

Theorems 1 and 2 proof that our approach mirrors exactly the semantics of annotated disjunctions into ProbLog programs with FOL constraints. Example 8 illustrates the generality of our approach by encoding ADs which have the same atoms in their heads, i.e. the same event can result from multiple causes.

Example 8. Consider again the same problem as in the previous examples: a bag with colored balls. According to one source of information in the bag there are red, green and blue balls (as in Example 2); another source states that there are only red and green balls in the bag. This knowledge is expressed by the following program:

r_1: 0.6::red(b1); 0.3::green(b1); 0.1::blue(b1)<-pick(b1).

r_2: 0.7::red(b1); 0.3::green(b1) <- pick(b1).

r_3: 0.6::pick(b1); 0.4::no_pick(b1) < − true.

The probability tree associated with these ADs is:

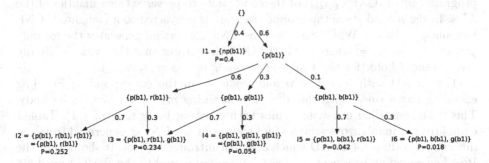

and the selections corresponding to the paths from the root to a leaf are:

Selection:	Interpretation	Probability $P(\sigma_i)$
σ_1	$I1 = \{no_pick(b1)\}$	0.4
σ_2	$I2 = \{pick(b1), red(b1)\}$	0.252
σ_3	$I3 = \{pick(b1), red(b1), green(b1)\}$	0.108
σ_4	$I3 = \{pick(b1), red(b1), green(b1)\}$	0.126
σ_5	$I4 = \{pick(b1), green(b1)\}$	0.054
σ_6	$I5 = \{pick(b1), red(b1), blue(b1)\}$	0.042
σ_7	$I6 = \{pick(b1), green(b1), blue(b1)\}$	0.018

Picking from the bag can result in selecting a red or green ball or red, green or blue ball (according to the different sources of information). In this case there are two selections (σ_2 and σ_3) which define the same interpretation ($I2$). When computing the MPE (as defined in Definition 2), each of these selections should be considered separately. The following table shows that our method corresponds exactly one possible world to each selections, regardless if they have the same interpretation and thus the MPE state can be computed correctly.

Possible World	pf(1, r)	pf(1, g)	pf(1, b)	pf(2, r)	pf(2, g)	pf(3, p)	pf(3, np)	r	g	b	p	np	Probability $P(\omega_i)$
ω_1	F	F	F	F	F	F	T	F	F	F	F	T	0.4
ω_2	T	F	F	T	F	T	F	T	F	F	T	F	0.252
ω_3	F	T	F	T	F	T	F	T	T	F	T	F	0.108
ω_4	F	F	T	T	F	T	F	T	F	T	T	F	0.126
ω_5	T	F	F	F	T	T	F	T	T	F	T	F	0.054
ω_6	F	T	F	F	T	T	F	F	T	F	T	F	0.042
ω_7	F	F	T	F	T	T	F	F	T	T	T	F	0.018

4   Evaluation

4.1   Analysis

The ProbLog pipeline for both MARG and MPE inferences consists of *Grounding*, *CNF conversion*, *sd-DNNF compilation* and *Evaluation* stages. Each previous stage produces input for the next one. ADs are processed in the Grounding stage. Using the ProbLog encoding the grounder generates only a ground program. On the other hand, invoking the WMC encoding results in (i) a ground program and (ii) CNF (φ_{ADs}) of the constraints to preserve the semantics of the ADs. In the second stage the ground program is converted to a (weighted) CNF formula φ_g. For the WMC encoding the CNF conversion generates the formula $\varphi = \varphi_g \wedge \varphi_{ADs} \wedge \varphi_E$, where φ_E is the CNF that states that the evidence (if any given) should hold (for the ProbLog encoding it is $\varphi = \varphi_g \wedge \varphi_E$).

For an AD with $|head| = n$ and $|body| = m$ the corresponding ProbLog encoding has n rules, with the nth rule containing $m + n - 1$ facts in its body. The WMC encoding constructs rules with constant body size ($m + 1$). Table 1 compares the number of variables and clauses in the CNF φ generated from the different encodings of an AD which does not introduce cycles. It shows that the ProbLog encoding produces smaller CNFs as compared to the WMC Encoding.

Both encodings deal with each AD independently and define one (ground ProbLog) rule for each dependency between a head atom and the probabilistic facts generated by the encoding. The bodies of rules with the same head are combined in a disjunction in the CNF and introduce additional variables and

Table 1. Size of generated CNFs by the ProbLog and the WMC encoding.

Encoding:	Number of CNF variables:	Number of CNF clauses:
ProbLog:	$2n + m - 1$	$\frac{n(2m+n+3)}{2} - 1$
WMC:	$2n + m$	$\frac{n(4m+n+3)}{2} + 1$

clauses. For the two encodings of a set of ADs which share head atoms the number of additional variables and clauses in the CNF is the same. ProbLog employs the algorithm of [21] to break cycles. It has the same impact on both encodings. Table 1 ignores the additional clauses and variables in the cases where the ADs share head atoms and the ones introduced during loop breaking as they are the same for both encodings.

The sd-DNNF compilation stage is computationally the hardest in the pipeline of ProbLog. We use c2d [22] to compile a CNF into a sd-DNNF. c2d is non-deterministic (cf. [23]), that is, for the same CNF the resulting sd-DNNFs may differ. Hence, we cannot make theoretical estimations on the time for generating an sd-DNNF, its size or the time for its evaluation.

Our implementation of the evaluation procedure for the MPE task is according to [16, Chapter12] and has the same complexity as the evaluation for the MARG task. Our experiments confirmed this claim.

4.2 Experimental Data

Our experiments aim to answer: (i) **What is the trade-off between the ProbLog encoding and the WMC encoding when doing MARG inference?**; (ii) **How does the WMC encoding scale w.r.t. the data size?**.

We analyze the results from experimenting on three artificially generated datasets. The first one (the *Balls* benchmark) is a more complex version of the ball game (cf. Example 2) that uses ADs to represent the different possible consequences of an action. The two other benchmarks are taken from [24]: we use the programs with annotated disjunctions with increasing head atoms (M_{gh}) and the ones with increasing number of (negated) body atoms (M_{gnb}). With the M_{gh} dataset we show the impact of the number of heads on the performance of the encodings. The M_{gnb} dataset shows the impact of the size of the bodies (that is, the second constraint of the WMC encoding). Our benchmarks can be found at http://people.cs.kuleuven.be/~dimitar.shterionov/mpe.

To assess the performance of our approach we measure the *processing-compilation* time (time for grounding plus CNF conversion plus sd-DNNF compilation), the evaluation time and the total inference time. We report the ratio $T^r = \frac{T(ProbLog)}{T(WMC)}$ per number of ground ADs which shows the relative time between the ProbLog and the WMC encoding. To assess the memory consumption we compare the number of nodes and edges for the generated sd-DNNF in each test run. We report the size ratio $S^r = \frac{S(ProbLog)}{S(WMC)}$.

Since c2d may generate different sd-DNNFs for the same CNF we run each test 5 times and report the average of time and size. We run our experiments on an 8-thread, Intel®Core™i7 @ 3400 MHz machine with 16GB RAM.

4.3 Experimental Results

Figure 1 summarizes the results for the *Balls* program. Figure 1 (a) shows the time ratio T^r w.r.t. the number of ADs and Fig. 1 (b) the size ratio S^r.

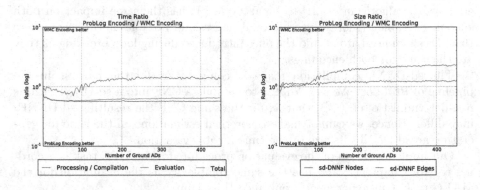

Fig. 1. Results from the *Balls* benchmark.

Figure 1 (a) shows a worse processing-compilation time (blue line) for the WMC encoding compared to the ProbLog encoding but better evaluation time (green line). Since the processing-compilation time is much larger than the evaluation, the total time (red line) is also worse for the WMC encoding. Furthermore, the total time for the case of the WMC encoding is higher (as compared to the ProbLog encoding) with a constant factor over the number of ground ADs.

From Fig. 1 (b) we conclude that for almost all queries the compiled sd-DNNFs for the WMC encoding are smaller. This, together with the better evaluation time shows the WMC encoding to be preferable, since e.g., in diagnostics typically the model is compiled once and evaluated many times.

The results from experimenting on the M_{gh} dataset, summarized in Fig. 2, show similar tendencies as the ones for the *Balls* benchmark. That is, the WMC encoding has worse processing-compilation and therefore also total execution time but still better evaluation time.

Crucial for the encoding is the size of the body of the AD as it influences drastically the size of the CNF (cf. Table 1). Figure 3 shows how this affects the execution time. For the M_{gnb} dataset the WMC encoding results in worse processing-compilation, total and evaluation time than the ProbLog encoding.

Analyzing the results of our experiments we can state that there is a trade-off between the ProbLog encoding and the WMC encoding: generally, the total

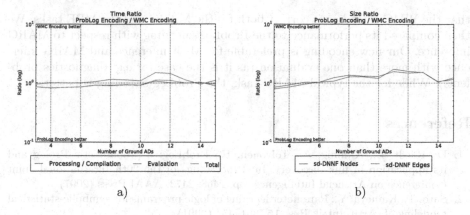

Fig. 2. Results from the M_{gh} benchmark.

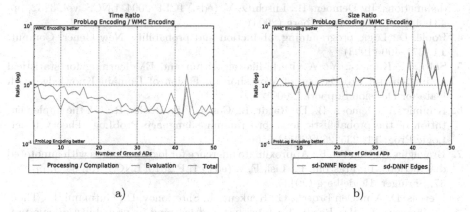

Fig. 3. Results from the M_{gnb} benchmark.

execution time for inference is worse (for the WMC encoding case). But the better evaluation time (*Balls* and M_{gh} benchmarks) due to the more compact sd-DNNFs makes our encoding preferable for problems where you compile once and evaluate many times. Also, in extreme cases, like the M_{gnb} benchmark, the WMC encoding does not perform better. Our experiments also show that with the two encodings ProbLog's inference scales in similar ways, with the ProbLog encoding outperforming the WMC encoding with respect to total execution time.

5 Conclusion

In this paper we introduced an encoding of annotated disjunctions as a weighted CNF formula (the *WMC Encoding*) to correctly reason with annotated disjunctions within the ProbLog system. ADs were supported previously by transforming them into probabilistic facts and Prolog rules. This approach (the *ProbLog encoding*) was correct for the MARG task but not for the MPE task. We showed that the WMC Encoding is correct both for the MARG and the MPE tasks. We then compared its performance to the ProbLog encoding with respect to MARG inference. Our new encoding is preferable for MPE inference and MARG inference with more than one evaluation (as it is the case for e.g. diagnostics problems). While for the typical MARG task, the ProbLog encoding is better.

References

1. De Raedt, L., Kimmig, A., Toivonen, H.: ProbLog: a probabilistic Prolog and its application in link discovery. In: Proceedings of the 20th International Joint Conference on Artificial Intelligence, pp. 2468–2473. AAAI Press (2007)
2. Sato, T., Kameya, Y.: Parameter learning of logic programs for symbolic-statistical modeling. J. Artif. Intell. Res. **15**, 391–454 (2001)
3. Vennekens, J., Verbaeten, S., Bruynooghe, M.: Logic programs with annotated disjunctions. In: Demoen, B., Lifschitz, V. (eds.) ICLP 2004. LNCS, vol. 3132, pp. 431–445. Springer, Heidelberg (2004)
4. Poole, D.: Logic programming, abduction and probability. New Gener. Comput. **11**, 377–400 (1993)
5. Sato, T., Kameya, Y.: A viterbi-like algorithm and EM learning for statistical abduction. In: Proceedings of Workshop on Fusion of Domain Knowledge with Data for Decision Support (2000)
6. Kimmig, A., Demoen, B., De Raedt, L., Costa, V.S., Rocha, R.: On the implementation of the probabilistic logic programming language ProbLog. Theory Pract. Logic Prog. **11**, 235–262 (2011)
7. Bragaglia, S., Riguzzi, F.: Approximate inference for logic programs with annotated disjunctions. In: Frasconi, P., Lisi, F.A. (eds.) ILP 2010. LNCS, vol. 6489, pp. 30–37. Springer, Heidelberg (2011)
8. Fierens, D., Van Den Broeck, G., Renkens, J., Shterionov, D., Gutmann, B., Thon, I., Janssens, G., De Raedt, L.: Inference and learning in probabilistic logic programs using weighted boolean formulas. Theor. Pract. Logic Program. (TPLP 2015) **15**(3), 358–401 (2015). doi:10.1017/S1471068414000076
9. Fierens, D., Van den Broeck, G., Thon, I., Gutmann, B., De Raedt, L.: Inference in probabilistic logic programs using weighted CNF's. In: Proceedings of the 27th Conference on Uncertainty in Artificial Intelligence, pp. 211–220 (2011)
10. Fierens, D., Van den Broeck, G., Bruynooghe, M., De Raedt, L.: Constraints for probabilistic logic programming. In: Roy, D., Mansinghka, V., Goodman, N., (eds.) Proceedings of the NIPS Probabilistic Programming Workshop (2012)
11. Sato, T.: A statistical learning method for logic programs with distribution semantics. In: Proceedings of the 12th International Conference on Logic Programming (ICLP95), pp. 715–729. MIT Press (1995)
12. Vennekens, J., Denecker, M., Bruynooghe, M.: CP-logic: A language of causal probabilistic events and its relation to logic programming. Theory Pract. Logic Prog. **9**, 245–308 (2009)
13. Gutmann, B.: On continuous distributions and parameter estimation in probabilistic logic programs. Ph.D thesis, KULeuven (2011)
14. Getoor, L., Taskar, B.: Introduction to Statistical Relational Learning (Adaptive Computation and Machine Learning). MIT Press, Cambridge (2007)
15. Koller, D., Friedman, N.: Probabilistic Graphical Models: Principles and Techniques. MIT Press, Cambridge (2009)

16. Darwiche, A.: Modeling and Reasoning with Bayesian Networks. Cambridge University Press, Cambridge (2009). Chapter 12
17. Taskar, B., Guestrin, C., Koller, D.: Max-margin markov networks. In: Proceedings of Neural Information Processing Systems (2003)
18. Richardson, M., Domingos, P.: Markov logic networks. Mach. Learn. **62**(1–2), 107–136 (2006)
19. Huang, J.: Solving map exactly by searching on compiled arithmetic circuits. In: Proceedings of the 21st National Conference on Artificial Intelligence (AAAI-06), pp. 143–148 (2006)
20. Meert, W.: Inference and learning for directed probabilistic logic models. Ph.D thesis, Informatics Section, Department of Computer Science, Faculty of Engineering Science. Blockeel, Hendrik (supervisor) (2011)
21. Mantadelis, T., Janssens, G.: Dedicated tabling for a probabilistic setting. In: Hermenegildo, M.V., Schaub, T. (eds.) ICLP (Tech. Communications). Vol. 7 of LIPIcs., Schloss Dagstuhl - Leibniz-Zentrum fuer Informatik, pp. 124–133 (2010)
22. Darwiche, A.: New advances in compiling CNF into decomposable negation normal form. In: Proceedings of the 16th European Conference on Artificial Intelligence, pp. 328–332 (2004)
23. Darwiche, A.: A compiler for deterministic, decomposable negation normal form. In: Dechter, R., Sutton, R.S. (eds.) AAAI/IAAI, pp. 627–634. AAAI Press/MIT Press (2002)
24. Meert, W., Struyf, J., Blockeel, H.: CP-logic theory inference with contextual variable elimination and comparison to BDD based inference methods. In: Proceedings of 19th International Conference of Inductive Logic Programming, pp. 96–109 (2009)

Towards Machine Learning of Predictive Models from Ecological Data

Alireza Tamaddoni-Nezhad[1]([⊠]), David Bohan[2], Alan Raybould[3], and Stephen Muggleton[1]

[1] Department of Computing, Imperial College London, London SW7 2AZ, UK
{a.tamaddoni-nezhad,s.muggleton}@imperial.ac.uk
[2] UMR 1347 Agroécologie, Pôle Ecoldur, BP 86510, 21065 Dijon Cedex, France
david.bohan@dijon.inra.fr
[3] Syngenta Crop Protection AG, Schwarzwaldallee 215, CH-4058 Basel, Switzerland
alan.raybould@syngenta.com

Abstract. In a previous paper we described a machine learning approach which was used to automatically generate food-webs from national-scale agricultural data. The learned food-webs in the previous study consist of hundreds of ground facts representing trophic links between individual species. These species food-webs can be used to explain the structure and dynamics of particular eco-systems, however, they cannot be directly used as general predictive models. In this paper we describe the first steps towards this generalisation and present initial results on (i) learning general functional food-webs (i.e. trophic links between functional groups of species) and (ii) meta-interpretive learning (MIL) of general predictive rules (e.g. about the effect of agricultural management). Experimental results suggest that functional food-webs have at least the same levels of predictive accuracies as species food-webs despite being much more compact. In this paper we also present initial experiments where predicate invention and recursive rule learning in MIL are used to learn food-webs as well as predictive rules directly from data.

1 Introduction

Machine Learning has previously been used in ecology (e.g. [5]), however, ecological data-mining is relatively a new emerging subject. For example large-scale ecological data from agricultural systems are nowadays being produced to evaluate the impacts of new technology, such as genetically modified crops. These large-scale data can be also used to develop models for predicting the effects of perturbation on agro-ecosystems.

We have recently demonstrated [12] that a logic-based machine learning method can be used to automatically generate plausible and testable food webs from ecological census data. Through a review of the literature, it was found that many of the learned trophic links are corroborated by the literature. In particular, links ascribed with high probability by machine learning are shown to correspond well with those having multiple references in the literature.

© Springer International Publishing Switzerland 2015
J. Davis and J. Ramon (Eds.): ILP 2014, LNAI 9046, pp. 154–167, 2015.
DOI: 10.1007/978-3-319-23708-4_11

In some cases novel, high probability links were suggested, and some of these have recently been tested and confirmed by subsequent empirical studies [13].

The learned species food-webs described in [12,13] consist of hundreds of ground facts (ground abductive hypotheses) representing trophic links between individual species. These food-webs can be used to explain the structure and dynamics of particular eco-systems. However, species-based food-webs cannot be directly used as general predictive models unless they are generalised or used together with general (i.e. non-ground) predictive models.

In this paper we describe the first steps towards this generalisation and present initial results on (i) learning general functional food-webs (i.e. trophic links between functional groups of species) and (ii) meta-interpretive learning of general predictive rules (e.g. about the effect of agricultural management).

2 Background and Related Work

To make good decisions about ecosystem management, e.g. the management of agricultural land for the optimal delivery of ecosystem services, it is necessary to have theories that predict the effects of perturbation on ecosystems. Network ecology, and in particular food-web approach, holds great promise as an approach to modeling and predicting the effects of perturbation on ecosystems. Networks of trophic links, also known as food-webs, which describe the flow of energy/biomass between species, are important for explaining ecosystem structure and dynamics. However, relatively few ecosystems have been studied through detailed food-webs because establishing predation relationships between the many hundreds of species in an ecosystem is expensive and in many cases impractical. This is mainly because establishing predation relationships between the many hundreds of species in an ecosystem is resource intensive, requiring considerable investment in field observation and laboratory experimentation.

We have recently developed [12] a logic-based machine learning method which can be used to automatically generate plausible and testable food-webs from ecological census data. The initial food-web was learned from an extensive Vortis suction sampling of invertebrates from 257 arable fields across the UK as part of the Farm Scale Evaluations (FSE) of genetically modified, herbicide-tolerant (GMHT) crops. Using a technique based on calculating a treatment effect ratio [6], this abundance count data was converted into up/down information and was regarded as the primary observational data for the learning. The set of observable (or training) data are represented by predicate $abundance(X, S, up)$ (or $abundance(X, S, down)$) expressing the fact that the abundance of X at site S is up (or $down$). This information was compiled from FSE data as detailed in [12]. The knowledge gap that we initially aimed to fill was a predation relationship between species. Thus, we declared abducible predicate $eats(X, Y)$ capturing the hypothesis that species X eats species Y. In order to use abduction, we also provided the rules which describe the observable predicate ($abundance$) in terms of the abducible predicate ($eats$):

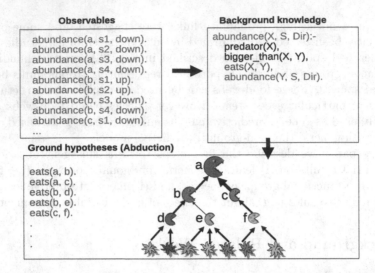

Fig. 1. Machine learning of species food-webs from ecological data using Abductive ILP.

$abundance(X,S,Dir):-$
 $predator(X),$
 $bigger\_than(X,Y),$
 $eats(X,Y),$
 $abundance(Y,S,Dir).$

where *Dir* can be either *up* or *down*. This Prolog rule expresses the inference that following a perturbation in the ecosystem (caused by the management), the increased (or decreased) abundance of species X at site S can be explained by X eating species Y, which is lower in the food chain and the abundance of species Y is increased (or decreased). It also includes additional conditions to constraint the search for abducible predicate $eats(X,Y)$, i.e. X should be a predator and bigger than Y. Predicates $predator(X)$ and $bigger\_than(X,Y)$ are provided as part of the background knowledge. Given this model and the observable data, the Abductive ILP system Progol 5 [10] was used to generate a set of ground abductive hypotheses in the form of 'eats' relations between species as shown in Fig. 1. These abductive hypotheses are automatically generated by matching the given information to the rule in order to abduce a fact which explains the observations. In this example, given the inputs, abduction will generate the hypotheses that a particular species a eats a particular species b. The set of ground hypotheses can be visualised as a network of trophic links (food-webs) as shown in Fig. 2a. In this network a ground fact $eats(a,b)$ is represented by a trophic link from species b to species a.

A probabilistic approach, called Hypothesis Frequency Estimation (HFE) [12], was used for estimating probabilities of hypothetical trophic links based on their frequency of occurrence when randomly sampling the hypothesis space. Using this technique, the thickness of trophic links in Fig. 2 represent

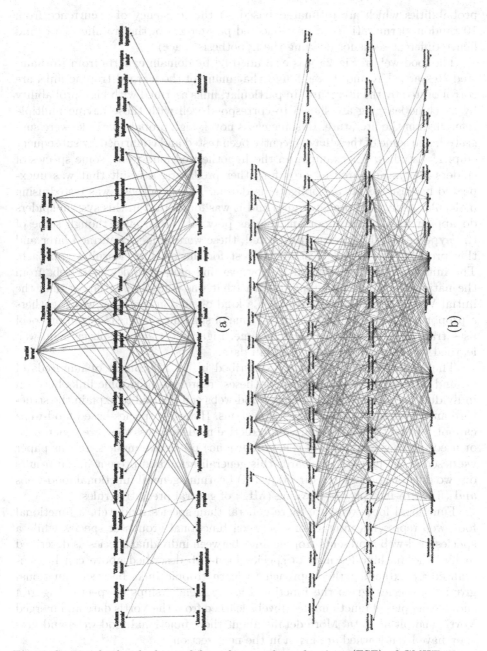

Fig. 2. Species food-webs learned from farm-scale evaluations (FSE) of GMHT crops data collected using Vortis and pitfall sampling methods from 257 fields across the UK. **(a)** Food-web learned from the Vortis data and **(b)** food-web learned from merged Vortis and pitfall data. Thickness of trophic links represents probabilities which are estimated using HFE [12].

probabilities which are estimated based on the frequency of occurrence from 10 random permutations (a user selected parameter) of the training data (and hence different seeds for defining the hypothesis space).

The food-web in Fig. 2a was examined [2] by domain experts from Rothamsted Research UK and it was found that many of the learned trophic links are corroborated by the literature. In particular, links ascribed with high probability by machine learning are shown to correspond well with those having multiple references in the literature. In some cases novel, high probability links were suggested, and some of these have recently been tested and confirmed by subsequent empirical studies. For example, in the hypothesised food-webs, some species of spiders always appeared as prey for other predators; a result that was unexpected because spiders are obligate predators. This hypothesis was tested using molecular analysis of predator guts and it was found that in this system spiders do appear to play an important role as prey [4]. Thus, even though some of the hypothesised links were unexpected, these were in fact confirmed later and this provided an extremely stringent test for the machine learning approach. The initial study was extended [13] by learning more complex food-webs from the national-scale pitfall sampling data (that was considerably larger than the initial Vortis data). Vortis protocol is a kind of suction sampling which gathers a proportion of species present at the time of sampling, whereas pitfall protocol uses trap sampling over a period of time. Figure 2b shows a species food-web learned from merged Vortis and pitfall data.

The learned species food-webs described in [12,13] consist of hundreds of ground facts (ground abductive hypotheses) representing trophic links between individual species (see Fig. 2). These food-webs can be used to explain the structure and dynamics of particular eco-systems. However, species-based food-webs cannot be directly used as general predictive models unless they are generalised or used together with general (i.e. non-ground) predictive models. In this paper we describe the first steps towards this generalisation and present initial results on two different but related directions: (i) Learning general functional food-webs and (ii) Meta-interpretive learning (MIL) of general predictive rules.

Functional food-webs are more general than species food-web, a functional food-web represents interactions between functional groups of species while a species food-web represents trophic links between individual species as described in [2]. The machine learning of species food-web described above can be generalised by extending the approach to learn trophic links between functional groups of species, given the functional group memberships of species. Figure 4 shows examples of functional food-webs learned from the Vortis data and merged Vortis and pitfall data. More details about these functional food-webs and how they have been learned are given in the next section.

The machine learning of species food-webs (and functional food-webs) described above assume that the logical rules describing the problem, e.g. a rule which describes the observable predicate in terms of 'eats' relations (or the functional group memberships of species) are given as background knowledge. However, this information may not be always available or it could be incomplete. In this paper we describe a new machine learning approach which allows automated discovery of trophic links as well as general predictive rules (and functional group memberships) directly from ecological data. This new setting requires both predicate invention and learning recursive rules which are not supported by most machine learning tools, including Progol which has been used for learning species and functional food-webs. On the other hand, Meta-interpretive learning (MIL) [9,11] is a new approach for predicate invention and recursive rule learning and can be used for learning ground hypotheses (e.g. trophic links) as well as non-ground hypotheses such as the general recursive rules.

The MIL setting was initially used [9] for learning grammars from example sequences but was extended [8] to dyadic definite clauses. Unlike some ILP systems which either support predicate invention or recursion learning, MIL was shown to be a very efficient approach for predicate invention as well as learning recursive programs. For example, the ILP system ATRE has been used [1] for the discovery of mutual recursive patterns from text. However, ATRE does not support invention of first-order predicates. MIL is also related to other studies where abduction has been used for predicate invention (e.g. [7]). One important feature of MIL, which distinguishes it from other existing approaches, is that it introduces new predicate symbols which represent relations rather than new objects or propositions.

3 Machine Learning of Predictive Models as an ILP Problem

The machine learning tasks described above, i.e. learning of species food-webs, learning of functional food-webs and learning of predictive rules, can all be formally described by adopting the general ILP setting. ILP systems use given set of positive and negative examples $E = \{E^+ \cup E^-\}$ and background knowledge B to construct a hypothesis H that explains E^+ relative to B such that the extended theory is self-consistent:

- $B \cup H \models E^+$, and
- $B \cup H \cup E^-$ is consistent.

The components E, B and H are each represented as logic programs. In the case of machine learning of species and functional food-webs, abductive learning is used to learn ground hypotheses H (abducible) in the form of *eats* relations between species or functional group of species. In this case, background knowledge includes general rules $R \subseteq B$ which describe the observable examples in terms of the abducible predicate (e.g. see definition of *abundance/3* above).

In the case of machine learning of predictive rules, Meta-interpretive learning (MIL) is used to learn a set of ground and non-ground hypotheses H. These include general predictive rules $R \subseteq H$ which describe the observable examples in terms of the invented predicates. In this case, background knowledge includes higher-order meta-rules $M \subseteq B$ which are activated during the proving of examples in order to generate hypotheses H.

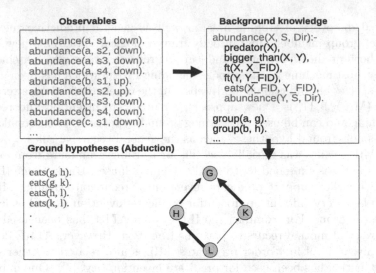

Fig. 3. Machine learning of functional food-webs from ecological data using Abductive ILP.

Hence, machine learning of species and functional food-webs only require abductive learning where predicates can be separated into two disjoint sets: the observable predicates and the abducible predicates. In practice, observable predicates describe the empirical observations of the domain, i.e. *abundance* of species. The abducible predicates describe underlying relations in our model, i.e. *eats* relations between species or functional group of species, that are not observable directly but can, through the theory B, bring about observable information.

By contrast, machine learning of predictive rules requires a combination of abduction and induction where the induction is needed to generate a set of non-ground hypotheses that contain universally quantified variables and can be used as general predictive rules. For this purpose, we use MIL which provides a tight integration of abduction and induction as described in [9].

In the following sections, machine learning of functional food-webs and meta-interepretive learning of predictive models are described with more details.

4   Machine Learning of Functional Food-Webs

In this section we explain how the approach for learning species food-webs has been extended for learning functional food-webs which are more general than species food-webs. We also show that functional food-webs can lead to higher predictive accuracy than species food-webs.

As discussed in Sect. 2, species food-webs can be used to explain the structure and dynamics of a particular eco-system. However, functional food-webs (i.e. which represent trophic interactions between functional groups of species) are more important for predicting changes in agroecosystem diversity and productivity [3]. Given the background information on functional type of each species,

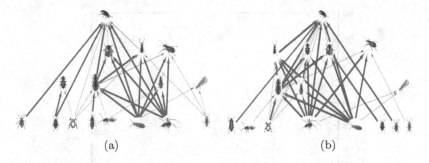

(a)   (b)

Fig. 4. Functional food-webs (a) learned from the Vortis data and (b) learned from merged Vortis and pitfall data. Each group in the functional food-web is represented by a species which can be viewed as an archetype for that functional group.

trophic networks for functional groups can be also learned from ecological data using the machine learning approach described above (See Fig. 3).

Here we need a rule which describes the observable predicate in terms of *eats* relation between functional groups:

abundance(X,S,Dir):-
    predator(X),
    bigger_than(X,Y),
    ft(X,X_FID),
    ft(Y,Y_FID),
    eats(X_FID,Y_FID),
    abundance(Y,S,Dir).

Given this new model and background information, e.g. functional types [3] of species in the form of $ft(X, X\_FID)$, trophic networks can be constructed for functional groups in a learning setting similar to the one described above for individual species.

Figure 4a and b show functional food-webs learned from the Vortis data and from merged Vortis and pitfall data respectively. These food-webs are constructed by learning trophic interactions between functional groups rather than individual species. Each functional group is represented by a species which can be viewed as an archetype for the functional group.

4.1 Empirical Evaluation

In this section we test the following null hypothesis:

Null hypothesis 1: A food-web constructed by learning trophic links between functional groups has a lower predictive accuracy compared to the food-web for individual species.

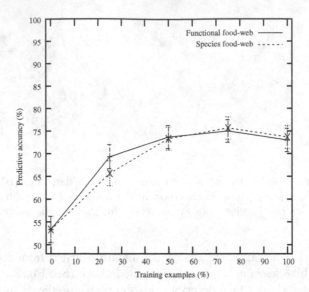

Fig. 5. Predictive accuracies of functional food-web vs. species food-web from leave-one-out cross-validation tests.

Materials and Methods. In this experiment Progol 5.0[1] is used to abduce 'eats' relations between species and functional groups of species from observable data (i.e. up/down abundance of species at different sites). The observable data has been compiled from FSE data as described in [12]. The up/down abundance of species at different sites are then encoded as predicates $abundance(X, S, up)$ and $abundance(X, S, down)$. The background knowledge includes information about sites and species and Prolog rules for $abundance$ as described in Sects. 2 and 4. Hypothesis Frequency Estimation (HFE) [12], was used for estimating probabilities of hypothetical trophic links based on the frequency of occurrence from 10 random permutations of the training data. Relative frequencies and probabilistic inference were also used to estimate probabilities of unseen data as described in [12]. For example, the probability $p(abundance(a, s, up))$ can be estimated by relative frequency of hypotheses which imply a at site s is up. Similarly, $p(abundance(a, s, down))$ can be estimated and by comparing these probabilities we predict whether the abundance of a at site s is up or $down$. This method used in the leave-one-out experiments in [12] to compare the predictive accuracies of probabilistic trophic networks vs non-probabilistic trophic networks. We use the same leave-one-out test strategy (described in Fig. 5 in [12]) to compare the predictive accuracies of species food-web vs functional food-web from Vortis data (food-webs shown in Figs. 2a and 4a). Other materials and methods are similar to those used in the experiments in [12], but we also include the rule and background knowledge for learning functional food-webs as described above.

---

[1] Available from: http://www.doc.ic.ac.uk/~shm/Software/progol5.0/.

Results and Discussion. The predictive accuracies of the functional and species food-webs are shown in Fig. 5. According to this figure, the difference between the predictive accuracies of the probabilistic network for the species food-web and the functional food-web are not significant when more than 50 % of training examples are provided. However, the predictive accuracy of functional food-web is significantly higher (p-value of 0.004 from t-test) than the predictive accuracy of species food-web when 25 % of training examples is available. This result suggests that when the number of training examples are limited, the functional food-web (which is more general) has a higher predictive accuracy compared to the species food-web. The null hypothesis 1 can therefore be refuted.

In general this experiment confirms that a food-web which is constructed by learning trophic links between functional groups is at least as accurate as the food-web for individual species despite being less complex (i.e. having less nodes and edges). Note that in this experiment we used a leave-one-out test strategy and evaluated both food-webs on the data from the same agricultural system. We expect the higher predictive accuracy of the functional food-web to be more evident if the food-webs are evaluated on a different agricultural system (e.g. different crops, climate etc.) where different species (not present in the training data) may exist. In this case we expect the species food-web to have a lower predictive accuracy and we intend to demonstrate this in a future work.

5   Meta-Interepretive Learning of Predictive Models

In the machine learning setting described in the previous section, the recursive rules describing the observable predicate (*abundance*/3) and the functional groups were provided as part of background knowledge. However, this information may not be always available or it could be incomplete. Here we describe a new machine learning setting where these information could be learned directly from data.

This new learning setting requires predicate invention and recursive rule learning and we use Meta-interpretive learning (MIL) [9,11] for this purpose. In the MIL framework described in this paper, predicate invention is conducted via construction of substitutions for meta-rules employed by a meta-interpreter. The use of the meta-rules clarifies the declarative bias being employed. New predicate names are introduced as higher-order skolem constants, a finite number of which are introduced during every iterative deepening of the search as described in [11]. Both predicate invention and the learning of recursion are implemented in MIL via metalogical substitutions with respect to a modified Prolog meta-interpreter which acts as the learning engine. The meta-interpreter is provided by the user with meta-rules which are higher-order expressions describing the forms of clauses permitted in hypothesised programs. The meta-interpreter attempts to prove the examples and, for any successful proof, saves the substitutions for existentially quantified variables found in the associated meta-rules. When these substitutions are applied to the meta-rules they result in a first-order definite program which is an inductive generalisation of the examples.

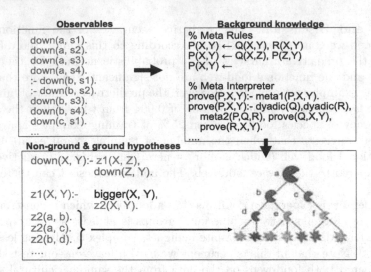

Fig. 6. Meta-interpretive learning of ground hypotheses (i.e. food-web) and non-ground hypotheses (i.e. prediction rule for down regulation) learned from a simplified food-web data (abundance of 6 species from 4 different sites). Predicates $z1$ and $z2$ are invented predicates, where $z2$ represents 'eats' relation.

Figure 6 shows meta-interpretive learning of ground hypotheses (i.e. food-web) and non-ground hypotheses (i.e. prediction rule for down regulation) learned from a simplified ecological data on down regulation of species following an agricultural management. MIL works by proving examples via meta-interpreter. This figure shows three higher-order meta-rules which are activated during the proof in order to generate the hypotheses shown in this figure. These hypotheses include non-ground rules and ground facts. Predicates $z1$ and $z2$ are invented predicates, where $z2$ represents 'eats' relation. Hence, the ground facts $z2(a, c)$, $z2(c, f)$, etc. represent the food-web which together with the non-ground rules for 'down' can be used for predicting down-regulation. The rule for 'down' shown in Fig. 6 is comparable to the rule provided as background knowledge in the previous sections.

5.1 Empirical Evaluation

In this section we test the following null hypothesis:

Null hypothesis 2: The MIL system *Metagol* cannot outperform the ILP system *Progol* in learning prediction rules as well as trophic links from a simplified ecological data.

Materials and Methods. In this section we use $Metagol_D{}^2$ to learn ground hypotheses (i.e. food-web) and non-ground hypotheses (i.e. prediction rule for

---

[2] Available from: http://ilp.doc.ic.ac.uk/metagolD_MLJ/.

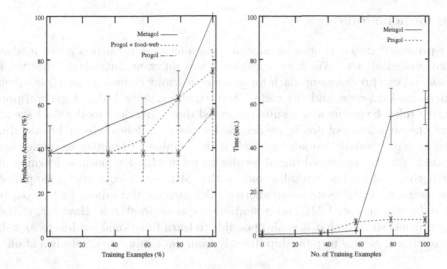

Fig. 7. Predictive accuracies and timing of Metagol vs Progol in learning ground hypotheses (i.e. food-web) and non-ground hypotheses (i.e. prediction rule for down regulation) from a simplified ecological data as shown in Fig. 6.

down regulation) from a simplified food-web data consisting of abundance of 6 species from 4 different sites, as shown in Fig. 6. We use a leave-one-out test strategy, similar to the one used in Sect. 4.1, to compare the predictive accuracies of *Metagol* vs *Progol*. *Progol* has been also tested in an enhanced mode where the food-web is provided as background knowledge (*Progol + foodweb*).

Results and Discussion. Figure 7 compares predictive accuracies of *Metagol* vs *Progol* vs *Progol + foodweb* in learning ground hypotheses (i.e. food-web) and non-ground hypotheses (i.e. prediction rule for down regulation) from the simplified food-web data described above. According to this, the predictive accuracies of *Metagol* are significantly higher than *Progol*. The accuracy of an enhanced Progol setting, where the food-web is provided as background knowledge (*Progol + foodweb*), reaches around 75 %. However, *Metagol*, which can learn both food-web and prediction rules, reaches an accuracy of 100 %. These results suggest that *Metagol* can learn the recursive rules and the food-web at the same time but it is difficult for Progol to learn these recursive rules directly from data even if the food-web structure is provided as background knowledge. Figure 7 also compares timings of *Metagol* vs *Progol*. According to this figure Progol is significantly faster. But it should be noted that unlike Progol which fails to learn any recursive rule, *Metagol* is learning and evaluating recursive rules.

6   Conclusions

We presented initial results on machine learning of general predictive models from ecological data. We have considered two different but related directions to extend our previous approach for machine learning of food-webs: (i) learning functional food-webs and (ii) meta-interpretive learning (MIL) of general predictive rules. Experimental results suggested that functional food-webs have at least the same levels of predictive accuracies as species food-webs and could also lead to higher predictive accuracy when the number of training examples are limited. We also presented initial results on using MIL for machine learning of predictive models. These results confirm that MIL can re-construct a simplified food-web and learn recursive predictive rules directly from data. In this paper we only demonstrated MIL on a simplified species food-web. However, initial experiments suggest that it is also possible to learn functional food-webs as well as functional groups membership directly from data using predicate invention.

Acknowledgements. The authors thank the members of Syngenta University Innovation Centre (UIC) at Imperial College, in particular Stuart Dunbar for his encouragement and support. We also thank Guy Woodward, Christian Mulder, Michael Traugott and Antonis Kakas for helpful discussions and support and the anonymous referees for useful comments. The first author acknowledges the support of an EPSRC "Pathways to Impact Award" during the writing of this paper.

References

1. Berardi, M., Malerba, D.: Learning recursive patterns for biomedical information extraction. In: Muggleton, S.H., Otero, R., Tamaddoni-Nezhad, A. (eds.) ILP 2006. LNCS (LNAI), vol. 4455, pp. 79–93. Springer, Heidelberg (2007)
2. Bohan, D.A., Caron-Lormier, G., Muggleton, S.H., Raybould, A., Tamaddoni-Nezhad, A.: Automated discovery of food webs from ecological data using logic-based machine learning. PLoS ONE **6**(12), e29028 (2011)
3. Caron-Lormier, G., Bohan, D.A., Hawes, C., Raybould, A., Haughton, A.J., Humphry, R.W.: How might we model an ecosystem? Ecol. Model. **220**(17), 1935–1949 (2009)
4. Davey, J., Vaughan, I., King, R.A., Bell, J., Bohan, D., Bruford, M., Holland, J., Symondson, W.: Intraguild predation in winter wheat: prey choice by a common epigeal carabid consuming spiders. J. Appl. Ecol. **50**(1), 271–279 (2013)
5. Dietterich, T.G.: Machine learning in ecosystem informatics and sustainability. In: Proceedings of the 21st International Joint Conference on Artificial Intelligence, IJCAI, Pasadena, Calif., pp. 8–13 (2009)
6. Haughton, A.J., Champion, G.T., Hawes, C., Heard, M.S., Brooks, D.R., Bohan, D.A., Clark, S.J., Dewar, A.M., Firbank, L.G., Osborne, J.L., et al.: Invertebrate responses to the management of genetically modified herbicide-tolerant and conventional spring crops. ii. within-field epigeal and aerial arthropods. Philos. Trans. R. Soc. Lond. Ser. B Biol. Sci. **358**(1439), 1863 (2003)
7. Inoue, K., Furukawa, K., Kobayashi, I., Nabeshima, H.: Discovering rules by meta-level abduction. In: De Raedt, L. (ed.) ILP 2009. LNCS, vol. 5989, pp. 49–64. Springer, Heidelberg (2010)

8. Muggleton, S.H., Lin, D.: Meta-interpretive learning of higher-order dyadic data-log: predicate invention revisited. In: Proceedings of the 23rd International Joint Conference Artificial Intelligence (IJCAI 2013), pp. 1551–1557 (2013)
9. Muggleton, S.H., Lin, D., Pahlavi, N., Tamaddoni-Nezhad, A.: Meta-interpretive learning: application to grammatical inference. Mach. Learn. **94**, 25–49 (2014)
10. Muggleton, S.H., Bryant, C.H.: Theory completion using inverse entailment. In: Cussens, J., Frisch, A.M. (eds.) ILP 2000. LNCS (LNAI), vol. 1866, p. 130. Springer, Heidelberg (2000)
11. Muggleton, S.H., Lin, D., Tamaddoni-Nezhad, A.: Meta-interpretive learning of higher-order dyadic datalog: predicate invention revisited. Mach. Learn. 100, 49–73 (2015). doi:10.1007/s10994-014-5471-y
12. Tamaddoni-Nezhad, A., Bohan, D., Raybould, A., Muggleton, S.H.: Machine learning a probabilistic network of ecological interactions. In: Muggleton, S.H., Tamaddoni-Nezhad, A., Lisi, F.A. (eds.) ILP 2011. LNCS, vol. 7207, pp. 332–346. Springer, Heidelberg (2012)
13. Tamaddoni-Nezhad, A., Milani, G., Raybould, A., Muggleton, S., Bohan, D.: Construction and validation of food-webs using logic-based machine learning and text-mining. Adv. Ecol. Res. **49**, 225–289 (2013)

PageRank, ProPPR, and Stochastic Logic Programs

Dries Van Daele$^{(\boxtimes)}$, Angelika Kimmig, and Luc De Raedt

Department of Computer Science, KU Leuven, Leuven, Belgium
{dries.vandaele,angelika.kimmig,luc.deraedt}@cs.kuleuven.be

Abstract. A key feature of ProPPR, a recent probabilistic logic language inspired by stochastic logic programs (SLPs), is its use of personalized PageRank for efficient inference. We adopt this view of probabilistic inference as a random walk over a graph constructed from a labeled logic program to investigate the relationship between these two languages, showing that the differences in semantics rule out direct, generally applicable translations between them.

1 Introduction

The need to combine reasoning about relational data and reasoning under uncertainty has led to a variety of languages and formalisms for statistical relational learning [5] and probabilistic logic learning and programming [4], but a deep understanding of the relation between different languages is often lacking. In this paper, we provide a detailed account of the relation between two such languages, namely stochastic logic programs (SLPs) [3,6], one of the oldest languages in the field, and the recently introduced probabilistic Prolog ProPPR [9,10], whose semantics is inspired by that of SLPs. Both languages use a labeled logic program to define a probability distribution over all answers to a query, but ProPPR biases this distribution towards answers obtained by shorter derivations in the program. This makes it possible to use variations of the PageRank algorithm [7], originally proposed for ranking web pages, for efficient probabilistic inference. We use this view of inference as a random walk over a graph constructed from a labeled logic program as the formal framework for our comparison, which shows that the differences in semantics rule out direct, generally applicable translations between programs in the two languages.

This paper is structured as follows. We review the background in Sect. 2. Section 3 focuses on the structure of the graphs on which the random walks are performed. Section 4 discusses the transition probabilities, highlighting their influence on approximate inference.

2 Background

We start by reviewing the background and formal notation.

© Springer International Publishing Switzerland 2015
J. Davis and J. Ramon (Eds.): ILP 2014, LNAI 9046, pp. 168–180, 2015.
DOI: 10.1007/978-3-319-23708-4_12

2.1   PageRank

The PageRank algorithm [7] was originally introduced for the purpose of ranking web pages on the World Wide Web. The *PageRank vector* is a probability distribution computed by applying PageRank on a graph. For each node in the graph, the PageRank vector specifies a probability that expresses its relative importance. Intuitively, this distribution can be regarded as the likelihood that a 'random surfer' [2] arrives at the respective web page. By regarding the World Wide Web as a graph where web pages are nodes and hyperlinks are edges, a 'random surfer' can be simulated by executing a random walk on it. In a random walk on an unweighted, directed graph, one moves from one node to the next by choosing uniformly between outgoing edges. More generally, we can consider a weighted, directed graph where a distribution is defined over the outgoing edges of each node. We can formalise this as a Markov chain whose states are the nodes in the graph, and whose transition matrix S is defined by the probabilities of moving from any node to another. The 'random surfer' model provides an extension to ensure that the random walk can reach every part of the graph. For this purpose, a jump to any node in the graph is introduced that occurs with probability $\gamma \in (0,1)$. This jump can be represented using a distribution s called the *personalization vector*. When s is not uniformly distributed over all nodes, the algorithm is referred to as personalized PageRank. Given the transition matrix A that captures the behaviour of the 'random surfer' model, i.e. $A = ((1-\gamma)S + \gamma\mathbf{1}s)$ with $\mathbf{1}$ a column vector of ones, the personalized PageRank vector π is defined as its stationary distribution:

$$\pi = A^T \pi$$

The PageRank vector can be computed using a simple iterative method called the power iteration method. Given an initial distribution x_0 over the nodes, the power iteration method computes $x_{k+1} = A^T x_k$ for increasing values of k, until the resulting vector has converged towards the PageRank vector. It thus simulates the flow of probability through the graph until a fixpoint is reached.

2.2   Logic Programming

A *definite clause program* $\mathcal{L}$ (or *logic program* for short) is a set of definite clauses. In a definite clause $a \leftarrow b$, the head a consists of a single atom, and the body $b = a_1, \ldots, a_n$ is a conjunction of atoms. A *substitution* θ is an expression of the form $\{V_1/t_1, \ldots, V_m/t_m\}$ where the V_i are different variables and the t_i are terms. Applying a substitution θ to a formula f results in the expression $f\theta$ where all variables V_i in f have been simultaneously replaced by their corresponding terms t_i in θ. Two formulas f_1 and f_2 can be *unified* if and only if there are substitutions θ_1 and θ_2 such that $f_1\theta_1 = f_2\theta_2$.

The main inference task in logic programming is to determine whether a given atom, also called *query* (or *goal*), is true in the least Herbrand model of a logic program. If the answer is yes (or no), we also say that the query *succeeds*

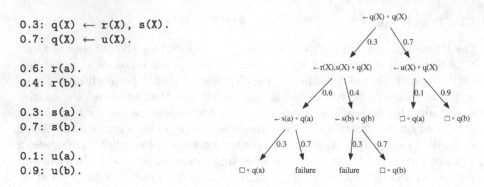

```
0.3: q(X) ← r(X), s(X).
0.7: q(X) ← u(X).

0.6: r(a).
0.4: r(b).

0.3: s(a).
0.7: s(b).

0.1: u(a).
0.9: u(b).
```

Fig. 1. Example of a pure and complete SLP and its SSLD tree for query $\leftarrow q(X)$

(or *fails*). If such a query is not ground, inference asks for the existence of an *answer substitution*, that is, a substitution that grounds the query into an atom that is part of the least Herbrand model.

Prolog answers queries using *refutation*, that is, the negation of the query is added to the program and resolution is used to derive the empty clause. This process, which continues until the empty goal is reached, can be depicted by means of an *SLD-tree*. The root of such a tree corresponds to the query, each branch to a *derivation*, that is, a sequence of clauses used for resolution steps. To simplify notation when discussing ProPPR later in the paper, the following definition labels each node in the SLD tree with both the subgoal at that node and the query as instantiated by the derivation leading to that node.

Definition 1 (SLD tree). *For a logic program $\mathcal{L}$ and query q, the SLD tree $\mathcal{T}$ is the labeled tree constructed as follows:*

- *The root of $\mathcal{T}$ is labeled $\langle \leftarrow q \circ q \rangle$.*
- *Each node $N = \langle \leftarrow q_1, \ldots, q_n \circ q\theta_N \rangle$ in $\mathcal{T}$ has a child node N_C for every clause $C = h \leftarrow b_1 \ldots, b_m$ in $\mathcal{L}$ whose head h has the same predicate as q_1. If h and q_1 cannot be unified, N_C is labeled with* failure, *and called a* failure *node. Else, let θ be the most general unifier of h and q_1. If the resolvent is the empty clause, N_C is labeled $\langle \Box \circ q\theta_N\theta \rangle$, and called a* success *node, else, it is labeled $\langle \leftarrow (b_1, \ldots, b_m, q_2, \ldots, q_n)\theta \circ q\theta_N\theta \rangle$.*

2.3 Stochastic Logic Program

A stochastic logic program (SLP) [3,6] is a definite clause program whose clauses have a probabilistic interpretation. A *pure* SLP labels each of its clauses with a probability (i.e. $p : a \leftarrow b$ where $p \in [0, 1]$). In a *complete* SLP, the probability labels of the clauses that share the predicate symbol in their head sum up to one. We use the pure and complete SLP in Fig. 1 as our running example.

Definition 2 (SSLD tree). *The stochastic SLD tree (SSLD tree) $\mathcal{T}_{\mathcal{S}}$ for a pure SLP $\mathcal{S}$ and query q is the SLD tree for q and $\mathcal{S}$, where each edge is labeled with*

```
about(X,Z) :- handLabeled(X,Z) # base.
about(X,Z) :- sim(X,Y),about(Y,Z) # prop.
sim(X,Y) :- links(X,Y) # sim,link.
sim(X,Y) :- hasWord(X,W),hasWord(Y,W),linkedBy(X,Y,W) # sim,word.
linkedBy(X,Y,W) :- true # by(W).
```

Fig. 2. Example ProPPR program from [9]; atoms to the right of # are rule features.

the probability of the clause used in the corresponding resolution step. $\mathcal{N}_\theta$ is the set of nodes labeled $\langle \Box \circ q\theta \rangle$, and $\mathcal{N}$ the union of all such $\mathcal{N}_\theta$. For each node $N \in \mathcal{N}$, we have

$$P_S(N) = \prod_{p_i:C_i \in d(N)} p_i$$

where $d(N) = p_1 : C_1, \ldots, p_n : C_n$ is the derivation ending in N.

In the SSLD tree for our example, cf. Figure 1, the probability of the leftmost derivation $0.3 : q(X) \leftarrow r(X), s(X), 0.6 : r(a), 0.3 : s(a)$ is $0.3 \cdot 0.6 \cdot 0.3 = 0.054$, and similar for all others. Note that several nodes can have the same label (e.g., two nodes are labeled $\langle \Box \circ q(a) \rangle$), so we cannot refer to a node by its label only.

Definition 3 (SLP probability P_S). *Given a pure SLP S, the probability of a query q with answer substitution θ is the sum of probabilities of all success nodes in $\mathcal{T}_S$ with label $\langle \Box \circ q\theta \rangle$, normalized for successful derivations:*

$$P_S(q\theta) = \frac{\sum_{N \in \mathcal{N}_\theta} P_S(N)}{\sum_{M \in \mathcal{N}} P_S(M)}$$

In our example, we get

$$P_S(q(a)) = (0.054 + 0.07)/(0.054 + 0.07 + 0.084 + 0.63) = 0.148$$
$$P_S(q(b)) = (0.084 + 0.63)/(0.054 + 0.07 + 0.084 + 0.63) = 0.852.$$

2.4 ProPPR

The probabilistic logic programming language ProPPR [9] is a recent language similar to SLPs, but with a bias towards shorter derivations, which allows for efficient approximate inference. ProPPR does not directly attach probabilities to its clauses. Instead, ProPPR supports a more flexible weighting of clauses through the use of weighted features. Each clause is annotated with one or multiple features, whose weights can be learned from data. An example of some ProPPR rules can be found in Fig. 2.

Here, we abstract away from the internal structure of the clause labels, and treat them as numerical weights. This abstraction is possible, because in essence, features associate a fixed, numerical weight with each instance of a clause after substitution with the most general unifier during SLD resolution. ProPPR gives special treatment to a set of facts called the database facts. All database facts have the same default weight w that is not modified during learning. Additionally, the semantics for predicates defined using these facts varies from those defined using normal ProPPR rules. Considering this, we can abstract away from the weighted features, but have to maintain a distinction between predicates defined using standard ProPPR rules, and those defined using database facts. We define a ProPPR program $\mathcal{P}$ as a tuple $(\mathcal{C}_p, \mathcal{C}_{db})$ where $\mathcal{C}_p$ is a set of weighted, definite clauses $w_i : C_i$, and $\mathcal{C}_{db}$ is a set of database facts f_j, all with the same default weight $w_j = w$.

Similar to how an SSLD tree is used to depict the process of deriving answers for a given SLP and query, the process employed by ProPPR can be captured in the PageRank ProPPR graph.

Definition 4 (PageRank ProPPR graph). *For ProPPR program $\mathcal{P}$, query q, restart weight $\alpha \in (0,1)$, and loop weight $\beta > 0$, the PageRank ProPPR graph $\mathcal{G}_{\mathcal{P},\alpha,\beta,q} = (\mathcal{N}_{\mathcal{G}}, \mathcal{E}_{\mathcal{G}})$ is defined as follows. Let $\mathcal{T}$ be the SLD tree for q and $\mathcal{P}$.*

1. *For each label $\langle \leftarrow s \circ q\theta \rangle$ or $\langle \square \circ q\theta \rangle$ appearing in $\mathcal{T}$, $\mathcal{N}_{\mathcal{G}}$ contains a unique node with that label.*
2. *$\mathcal{E}_{\mathcal{G}}$ contains an edge (l_1, l_2) if there is an edge between a pair of nodes labeled l_1 and l_2 in $\mathcal{T}$. The edge is labeled with the weight of the clause applied in the corresponding resolution step.*
3. *For every node $N \in \mathcal{N}_{\mathcal{G}}$, $\mathcal{E}_{\mathcal{G}}$ contains a restart edge $(N, \langle \leftarrow q \circ q \rangle)$ to the query node with label α.*
4. *For every success node $N = \langle \square \circ q\theta \rangle$, $\mathcal{E}_{\mathcal{G}}$ contains an edge (N, N) with label β.*

The PageRank ProPPR graph thus does not contain failure nodes. As node labels are unique, we also refer to a node by its label. Figure 3 shows the PageRank ProPPR graph derived from the SSLD tree in Fig. 1. In contrast to the labels in an SSLD tree, the labels in a PageRank ProPPR graph do not directly correspond to the transition probabilities of a random walk. ProPPR normalizes these labels to obtain transition probabilities, using a different normalization strategy depending on whether the predicate associated with the node in question is defined in $\mathcal{C}_p$ or $\mathcal{C}_{db}$. If the predicate is defined in $\mathcal{C}_p$, then the distribution over the outgoing edges is determined by normalizing over all outgoing edges.

If it is a database fact defined in $\mathcal{C}_{db}$, the weight of the restart edge remains unchanged, and the remaining outgoing edges are normalized to add up to $1 - \alpha$. Figure 4 shows the PageRank ProPPR graph in Fig. 3 with edges labeled with the transition probabilities obtained through normalization. Note that all the predicates in the example are defined in $\mathcal{C}_p$.

The loop weight β for the PageRank ProPPR graph in Fig. 3 is set to 1. This ensures that after normalization, the restart probability in the success node is similar to that of the other nodes in the graph.

Definition 5 (PageRank ProPPR distribution). *A ProPPR program $\mathcal{P}$ with weights α and β defines a distribution over all answer substitutions θ for a*

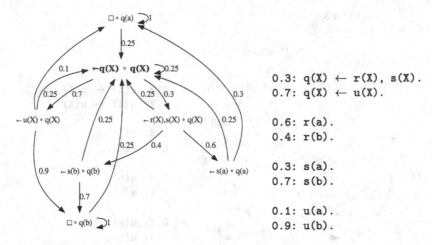

0.3: q(X) ← r(X), s(X).
0.7: q(X) ← u(X).

0.6: r(a).
0.4: r(b).

0.3: s(a).
0.7: s(b).

0.1: u(a).
0.9: u(b).

Fig. 3. PageRank ProPPR graph with restart weight $\alpha = 0.25$ and loop weight $\beta = 1$ derived from the SSLD tree in Fig. 1, with its corresponding program.

query q as follows. Let

$$\langle \Box \circ q\theta_1 \rangle, \ldots, \langle \Box \circ q\theta_k \rangle, \langle \leftarrow s_{k+1} \circ q\theta_{k+1} \rangle, \ldots, \langle \leftarrow s_m \circ q\theta_m \rangle$$

be an enumeration of all nodes in the PageRank ProPPR graph $\mathcal{G}_{\mathcal{P},\alpha,\beta,q}$, and $\pi = (\pi_1, \ldots, \pi_m)$ the PageRank vector defined by the graph. The PageRank ProPPR distribution is given by the vector $P_{\mathcal{P},\alpha,\beta}(q) = (P_1, \ldots, P_k)$ whose entries correspond to the normalized weight of the success nodes, i.e.,

$$P_i = \frac{\pi_i}{\sum_{j=1}^{k} \pi_j}$$

The PageRank ProPPR distribution for the example in Fig. 4 is (P_a, P_b) with

$$P_a = \pi_{\langle \Box \circ q(a) \rangle} / (\pi_{\langle \Box \circ q(a) \rangle} + \pi_{\langle \Box \circ q(b) \rangle}) = 0.19$$
$$P_b = \pi_{\langle \Box \circ q(b) \rangle} / (\pi_{\langle \Box \circ q(a) \rangle} + \pi_{\langle \Box \circ q(b) \rangle}) = 0.81.$$

3 Personalized PageRank over the SSLD Tree

Compared to the SSLD tree, the PageRank ProPPR graph adds restart edges on all nodes, adds self-loops on leaves, and omits failure nodes. We now discuss the effect of performing personalized PageRank over the SSLD tree with (a) both restart edges and self-loops added, and (b) only restart edges added. The discussion of failure nodes is deferred to Sect. 4.

As in the case of ProPPR, the additional edges specify the possible jumps, which allows us to incorporate the jump directly into the transition probabilities instead of using a personalization vector.

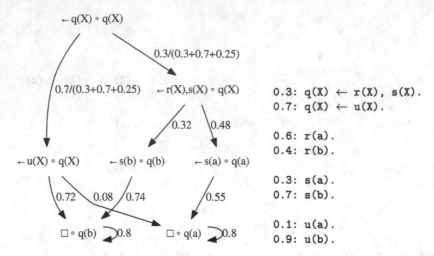

Fig. 4. The ProPPR example of Fig. 3 with transition probabilities as edge labels, omitting restart edges (labeled with the remaining probability mass at each node).

Definition 6 (PageRank SSLD graphs). *For SLP S, query q and restart probability $\gamma \in [0,1]$, the non-loopy PageRank SSLD graph $\mathcal{G}_{S,\gamma,q}$ consists of the nodes and edges of the SSLD tree $\mathcal{T}_S$ and an additional edge from every node to the root. An edge labeled p_i in $\mathcal{T}_S$ is labeled $p_i \cdot (1-\gamma)$ in $\mathcal{G}_{S,\gamma,q}$, edges from leaves to the root are labeled 1, and all other restart edges are labeled γ. The loopy PageRank SSLD graph $\mathcal{G}_{S,\gamma,q}^{\text{loop}}$ is $\mathcal{G}_{S,\gamma,q}$ extended with a direct self-loop for every leaf ($\langle \Box \circ q\theta \rangle$ or failure) of $\mathcal{T}_S$. An edge labeled p_i in $\mathcal{T}_S$ is labeled $p_i \cdot (1-\gamma)$ in $\mathcal{G}_{S,\gamma,q}^{\text{loop}}$, self-loops are labeled $1-\gamma$, and all restart edges are labeled γ.*

Note that all edge labels take values in [0,1] here, and for every node, labels of outgoing edges sum to 1, i.e., edge labels directly correspond to transition probabilities. Figure 5 shows the loopy PageRank SSLD graph for the SLP and query in Fig. 1 with $\gamma = 0.2$, where we omit restart edges to avoid clutter.

Definition 7 (PageRank SSLD distribution). *A (loopy or non-loopy) PageRank SSLD graph $\mathcal{G}$ defines a PageRank vector π with an entry π_N for every node N in the graph. For a query q with answer substitution θ, we obtain the PageRank SSLD probability*

$$P_{\mathcal{G}}(q\theta) = \frac{\sum_{N \in \mathcal{N}_\theta} \pi_N}{\sum_{M \in \mathcal{N}} \pi_M}$$

where $\mathcal{N}_\theta$ is the set of success nodes with answer substitution θ, and $\mathcal{N}$ the set of all success nodes.

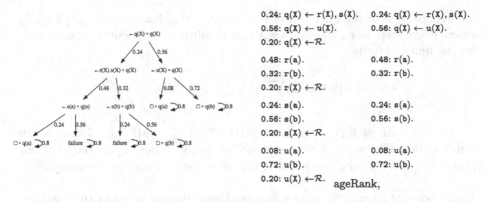

0.24: q(X) ← r(X), s(X).   0.24: q(X) ← r(X), s(X).
0.56: q(X) ← u(X).   0.56: q(X) ← u(X).
0.20: q(X) ←𝓡.

0.48: r(a).   0.48: r(a).
0.32: r(b).   0.32: r(b).
0.20: r(X) ←𝓡.

0.24: s(a).   0.24: s(a).
0.56: s(b).   0.56: s(b).
0.20: s(X) ←𝓡.

0.08: u(a).   0.08: u(a).
0.72: u(b).   0.72: u(b).
0.20: u(X) ←𝓡.   ageRank,

Fig. 5. The loopy PageRank SSLD graph (restart edges omitted) with restart probability 0.2 for the SLP in Fig. 1, an explicit notation for that SLP under the PageRank interpretation (middle) and the corresponding incomplete SLP (right).

For our example SLP S in Fig. 1, we have

$$P_{\mathcal{G}}(q(a)) = \frac{0.027648 + 0.0448}{0.027648 + 0.0448 + 0.4032 + 0.043008} = 0.1397$$

$$P_{\mathcal{G}}(q(b)) = \frac{0.4032 + 0.043008}{0.027648 + 0.0448 + 0.4032 + 0.043008} = 0.8603$$

These differ from the SLP probabilities $P_S(q(a)) = 0.148$ and $P_S(q(b)) = 0.852$, but can be obtained from a modified, incomplete SLP that scales the probabilities in S to account for the restart (shown on the right of Fig. 5 for our example).

Theorem 1. *For a given SLP S, query q and restart probability γ, let S_γ be the incomplete SLP obtained by multiplying all labels in S with $1 - \gamma$. For every answer substitution θ, we have*

$$P_{\mathcal{G}_{S,\gamma,q}}(q\theta) = P_{\mathcal{G}^{loop}_{S,\gamma,q}}(q\theta) = P_{S_\gamma}(q\theta).$$

We prove the theorem by showing that PageRank converges to a vector that defines the same normalized distribution over success nodes as S_γ.

Convergence It suffices to show that the PageRank SSLD Markov chain is irreducible, aperiodic, and positive-recurrent [8]. Since every state can be reached from and has a restart transition back to the query state, the chain is irreducible. Since the restart transition in the query node creates a loop, any state can be visited at an irregular time, and the chain is thus aperiodic.

Given that the chain is irreducible, positive recurrency directly follows for finite state spaces (as given by finite SSLD trees), and can be shown for infinite state spaces by showing that the mean recurrence time of one state (i.e. the expected number of steps before that state is first revisited) is finite. The mean recurrence time of the query state in the loopy PageRank SSLD graph $\mathcal{G}_{S,\gamma,q}^{\text{loop}}$, where every state has a transition with probability γ to the query state, is given by the finite quantity

$$\text{MRT}(q) = \sum_{n=1}^{\infty} n \cdot \gamma \cdot (1 - \gamma)^{n-1} = \frac{1}{\gamma}.$$

For the case of $\mathcal{G}_{S,\gamma,q}$, we have $\text{MRT}^{\text{loop}}(q) \leq \int_{\gamma} \text{MRT}(q)$. Let a be the infimum over the set of all restart transition probabilities in the Markov chain. Since $\int_{\gamma=a}^{1} \frac{1}{\gamma}$ for fixed $a > 0$ is finite, the query node is positive recurrent.

Equivalence of Distributions over Success Nodes We first consider the non-loopy graph $\mathcal{G}_{S,\gamma,q}$. Let π_q be the value associated with the query node in the corresponding PageRank vector π, and π_N the value associated with the success node reached by the SSLD refutation $p_1 : C_1, \ldots, p_n : C_n$, where the p_i are the labels in S. The corresponding labels in S_γ are thus $p_i' = p_i \cdot (1 - \gamma)$. We have

$$\pi_N = \pi_q \cdot \prod_{i=1}^{n} p_i'$$

which is the probability of the corresponding refutation in S_γ multiplied by the constant π_q. When normalizing over all successful derivations, this constant cancels out, and we thus obtain the same distribution as for S_γ. Similarly, for the loopy graph $\mathcal{G}_{S,\gamma,q}^{\text{loop}}$, we have

$$\pi_N = \pi_q \cdot \prod_{i=1}^{n} p_i' + (1 - \gamma) \cdot \pi_N = \frac{\pi_q}{\gamma} \cdot \prod_{i=1}^{n} p_i'$$

which again only differs by a multiplicative constant from the probability of the corresponding refutation in S_γ.   □

The effect of the restart can also be described by the logic program labeled with PageRank SSLD probabilities in the middle of Fig. 5. The special symbol $\mathcal{R}$ for the jump to the query node highlights the extra-logical nature of the restart step, which is similar to Prolog's cut operator !/0, but has a far greater effect.

Prolog implicitly builds an SLD tree by applying SLD resolution in a depth-first manner. Given a clause $b \leftarrow a_1, \ldots, a_m, !, \ldots, a_n$ Prolog commits to the clauses selected during the application of SLD resolution on the literals $b, a_1, \ldots, a_m$ when reaching the cut, i.e., Prolog does not consider any alternative clauses. This is equivalent to building the SLD tree for a program where cuts have been omitted, and then pruning away all alternative branches starting at the nodes associated with these literals. This pruning causes the search space to be reduced and results in the potential loss of solutions. Instead of directly pruning away alternative clauses, the restart clause causes the query to be stochastically restarted. Resolution on a goal such as $(\leftarrow \text{r}(X), \text{s}(X))$ using the restart clause for the first subgoal $\text{r}(X)$ results in the initial query with fresh variables, i.e., all subsequent subgoals (here just $s(X)$) are dropped, and all substitutions obtained so far forgotten. The restart affects the semantics by reducing the probability associated with all clauses for its predicate. When considering approximate inference (see Sect. 4) where branches are pruned when their associated probability drops below a particular threshold, we observe that this also results in branches being pruned earlier.

4   Discussion

We now discuss the differences between the transition probabilities used by PageRank SSLD and PageRank ProPPR, respectively. The types of nodes we need to consider are success nodes labeled $\langle \Box \circ q\theta \rangle$, failure nodes, and inner nodes labeled $N = \langle \leftarrow g_1, \ldots, g_m \circ q\theta \rangle$ with the predicate of g_1 defined in either $\mathcal{C}_p$ or $\mathcal{C}_{db}$. For such nodes N, let $\{w_1 : C_1, \ldots, w_k : C_k\}$ be all clauses in the program for which the head unifies with g_1, and $\{w_{k+1} : C_{k+1}, \ldots, w_n : C_n\}$ all those whose head has the same predicate as g_1, but does not unify with g_1.

For PageRank SSLD, $\sum_{i=1}^{n} w_i = 1.0$, and there is no distinction between $\mathcal{C}_p$ and $\mathcal{C}_{db}$. Based on Theorem 1, we only consider the loopy case here. Both success and failure nodes then have a restart probability of γ, and a self-loop probability of $1 - \gamma$. For an inner node N, the transition probabilities are γ for the restart edge, and $(1 - \gamma) \cdot w_i$ for the edge using clause C_i.

It is easy to verify that the same transition probabilities would be achieved by normalization over all outgoing edges with initial weight $\alpha = \frac{\gamma}{1-\gamma}$ on the restart edge, and $\beta = \sum_{i=1}^{n} w_i = 1$ on the self-loops. As our PageRank SSLD example uses $\gamma = 0.2$, we set $\alpha = 0.2/0.8 = 0.25$ and $\beta = 1$ in the corresponding PageRank ProPPR example. In the absence of failure edges ($n = k$ for all inner nodes), and with $\mathcal{C}_{db} = \{\}$ (as in SLPs), this choice of restart and self-loop weights makes the two semantics coincide.

In general, however, the transition probabilities of PageRank ProPPR depend on the failure nodes in the SSLD tree, as in inner nodes, the normalization does not include the weights of clauses for which unification fails.

If the first subgoal g_1 of inner node N is defined in $\mathcal{C}_{db}$, the restart probability in N is α, and the transition probability for each clause C_i with $1 \leq i \leq k$ is $p_i^p = (1 - \alpha)/k$, whereas the latter would be $p_i^s = (1 - \alpha)/n$ if normalized over all clauses as in PageRank SSLD. Thus, we have $p_i^p = p_i^s \cdot (n/k)$.

In the case of $\mathcal{C}_p$, the restart probability is $\alpha/(\alpha + \sum_{j=1}^{k} w_j)$, and the transition probability of such a C_i's edge is $p_i^p = w_i/(\alpha + \sum_{j=1}^{k} w_j)$. That is, the transition probability for a clause's edge increases as competing clauses fail. This effect can be quantified as follows:

$$p_i^p = \frac{\alpha + \sum_{j=1}^{k} w_j + \sum_{j=k+1}^{n} w_j}{\alpha + \sum_{j=1}^{k} w_j} \cdot p_i^s = \left(1 + \frac{\sum_{C_j \text{fails}} w_j}{\alpha + \sum_{C_j \text{suceeds}} w_j}\right) \cdot p_i^s$$

where $p_i^s = w_i/(\alpha + \sum_{j=1}^{k} w_j + \sum_{j=k+1}^{n} w_j)$ is the transition probability with normalization over all clauses as in PageRank SSLD.

As an illustration, consider the node labeled $\langle \leftarrow s(a) \circ q(a) \rangle$ in our running example. In the PageRank SSLD distribution defined by the program in Fig. 5, the probability of the edge associated with clause $s(a)$ is 0.24. In the PageRank ProPPR distribution in Fig. 4, the probability of this edge is $0.55 = (1 + \frac{0.7}{0.3+0.25})$. 0.24. Similarly, the probability of the edge below node $\langle \leftarrow s(b) \circ q(b) \rangle$ increases from 0.56 to 0.74. This also illustrates that the weights in the ProPPR program cannot directly be interpreted as the relative importance of different answers to a predicate, as answers with lower weight in the program may get a higher increase in transition probability if other clauses fail.

To summarize, each transition probability in the Markov chain given by a ProPPR program depends not only on the weight of the associated clause and the restart weight α, but also on the weight of the failing competing clauses. Because of this additional dependency, it is not possible to concisely represent a ProPPR program as an SLP. On the other hand, for a complete SLP whose SSLD tree contains no failure edges, the PageRank SSLD distribution with restart probability γ coincides with the PageRank ProPPR distribution with restart weight $\alpha = \frac{\gamma}{1-\gamma}$ and self-loop weight $\beta = 1$. In general, however, the different approaches to derive transition probabilities lead to different distributions.

The different restart probabilities of the nodes in the PageRank ProPPR graph also play a role in ProPPR's approximate inference algorithm [9], which is derived from the method for approximating personalized PageRank vectors used in the PageRank-Nibble algorithm [1]. This method assumes the personalized PageRank setup with transition matrix S, fixed jump probability γ and personalization vector s as discussed in Sect. 2.1. We discuss the implications in the context of ProPPR below, but first summarize the approach itself.

The algorithm has a stopping parameter $\epsilon \in (0, 1]$ and maintains two vectors p (the approximation of the PageRank vector) and r (the residual vector), each with an entry for every node in the graph. Initially, all entries in p are zero, and r is a distribution over nodes, which in the cases of interest to our discussion puts the full mass on the query node. Let $n(N)$ be the number of outgoing edges of node N, and p_i the transition probability from N to N_i. While there is a node N with $r(N) \geq \epsilon \cdot n(N)$, for one such N, the following updates are performed:

$$p(N) = p(N) + \gamma \cdot r(N)$$
$$r(N_i) = r(N_i) + (1 - \gamma) \cdot p_i \cdot r(N)$$
$$r(N) = 0$$

In time $O(\frac{1}{\gamma \cdot \epsilon})$, the algorithm converges to an approximation p of the PageRank vector where the error for every node N is bounded by $\epsilon \cdot n(N)$, considering no more than $\frac{1}{\gamma \cdot \epsilon}$ transition edges, and thus has a complexity independent of the size of the full graph (and its underlying program). Furthermore, this approach allows to (implicitly) construct the graph on the fly, expanding the vectors to new nodes as they first appear in the updates.

For PageRank SSLD, this algorithm can directly be applied with the loopy graph, by setting s to the vector with all mass on the query node, and S to the transition matrix with self-loop transition probabilities of 1 and the transition probabilities for all edges as given in the SSLD tree. Because of the tree structure (with loops on the leaves only) underlying the transition matrix, the stopping criterion corresponds to pruning the SSLD tree below nodes N with $P_{S_\gamma}(N) < \epsilon \cdot n(N)$ with $n(N)$ the total number of outgoing edges (including the restart).

For PageRank ProPPR, constructing the transition matrix and restart vector is more involved, as there is no constant restart probability γ, and furthermore, the values of restart probabilities are not known a priori. ProPPR therefore uses a constant γ that is expected to provide a lower bound on these values as the jump probability, sets s to the vector with all mass on the query node as in the SSLD case, and modifies the transition matrix S discussed for the SSLD case to also take into account the difference between the actual restart probabilities and the lower bound γ. In contrast to the SSLD case, this transition matrix corresponds to a graph with longer loops, where the stopping criterion is not directly expressible in terms of the derivation probability. It does however allow ProPPR to benefit from the advantages of the approximation method.

5 Conclusions

We have adopted ProPPR's view of probabilistic inference as a random walk over a graph constructed from a labeled logic program to investigate the relationship between ProPPR and SLPs. We have shown that the distribution obtained from a random walk with restart over the SSLD tree of a pure and complete SLP coincides with the distribution defined by an incomplete SLP with the same clauses, but with labels rescaled to take into account the restart probability. We have further shown that the differences in semantics rule out direct, generally applicable translations between ProPPR programs and SLPs.

Acknowledgements. Dries Van Daele is supported by PF/10/010 (NATAR). A. Kimmig is supported by the Research Foundation Flanders (FWO).

References

1. Andersen, R., Chung, F., Lang, K.: Using PageRank to locally partition a graph. Internet Math. **4**(1), 35–64 (2007)
2. Chebolu, P., Melsted, P.: PageRank and the random surfer model. In: Proceedings of the 19th Annual ACM-SIAM Symposium on Discrete algorithms (2008)
3. Cussens, J.: Stochastic logic programs: sampling, inference and applications. In: Proceedings of the 16th Conference on Uncertainty in Artificial Intelligence (2000)
4. De Raedt, L., Kersting, K.: Probabilistic Inductive Logic Programming. In: De Raedt, L., Frasconi, P., Kersting, K., Muggleton, S.H. (eds.) Probabilistic Inductive Logic Programming. LNCS (LNAI), vol. 4911, pp. 1–27. Springer, Heidelberg (2008)
5. Getoor, L., Taskar, B. (eds.): An Introduction to Statistical Relational Learning. MIT Press, Cambridge (2007)

6. Muggleton, S.: Stochastic logic programs. Adv. inductive logic prog. **32**, 254–264 (1996)
7. Page, L., Brin, S., Motwani, R., Winograd, T.: The PageRank citation ranking: bringing order to the web. Technical Report 1999–66, Stanford InfoLab, November 1999. Previous number = SIDL-WP-1999-0120
8. Resnick, S.I.: Adventures in Stochastic Processes. Birkhauser Verlag, Basel, Switzerland (1992)
9. Wang, W.Y., Mazaitis, K., Cohen, W.W.: Programming with personalized PageRank: a locally groundable first-order probabilistic logic. In: Proceedings of the 22nd ACM International Conference on information & knowledge management (2013)
10. Wang, W.Y., Mazaitis, K., Cohen, W.W.: ProPPR: Efficient first-order probabilistic logic programming for structure discovery, parameter learning, and scalable inference. In: AAAI Workshop on Statistical Relational Artificial Intelligence (StaRAI 2014) (2014)

Complex Aggregates over Clusters of Elements

Celine Vens[1,2,3]($\boxtimes$), Sofie Van Gassen[4], Tom Dhaene[4], and Yvan Saeys[1,2]

[1] Department of Respiratory Medicine, Ghent University, Ghent, Belgium
[2] VIB Inflammation Research Center, Ghent, Belgium
Yvan.Saeys@irc.vib-ugent.be
[3] Department of Public Health and Primary Care,
KU Leuven Kulak, Kortrijk, Belgium
Celine.Vens@kuleuven-kulak.be
[4] Department of Information Technology (INTEC)-iMinds,
Ghent University, Ghent, Belgium
{Sofie.VanGassen,Tom.Dhaene}@intec.ugent.be

Abstract. Complex aggregates have been proposed as a way to bridge the gap between approaches that handle sets by imposing conditions on specific elements, and approaches that handle them by imposing conditions on aggregated values. A complex aggregate summarises a subset of the elements in a set, where this subset is defined by conditions on the attribute values. In this paper, we present a new type of complex aggregate, where this subset is defined to be a cluster of the set. This is useful if subsets that are relevant for the task at hand are difficult to describe in terms of attribute conditions. This work is motivated from the analysis of flow cytometry data, where the sets are cells, and the subsets are cell populations. We describe two approaches to aggregate over clusters on an abstract level, and validate one of them empirically, motivating future research in this direction.

Keywords: Relational learning · Aggregation · Clustering · Flow cytometry

1 Introduction

In relational learning, examples are described by data residing in multiple tables of a relational database, linked together via one-to-one, one-to-many or many-to-many relationships. The latter two types of relationships result in single examples being related to a *set* of objects, and require mechanisms to deal with such sets. While, traditionally, relational learning approaches handled sets by either looking at properties of individual elements or by aggregating over them, a number of researchers [12,14,17,20,21] have looked into the combination of these approaches, resulting in so-called complex aggregates. Such complex aggregates aggregate over a subset of the elements of a set, where the subset is usually defined by imposing conditions on some attribute values. For instance, consider the widely used Mutagenesis dataset [18] of molecules that are classified as mutagenic or not, and are described by their atoms and bonds. Thus, any molecule

© Springer International Publishing Switzerland 2015
J. Davis and J. Ramon (Eds.): ILP 2014, LNAI 9046, pp. 181–193, 2015.
DOI: 10.1007/978-3-319-23708-4_13

is related to a *set* of atoms and bonds. A complex aggregate in this context is the number of carbon atoms of a molecule: it consists of an aggregation (namely, count) over a subset of the atoms (namely, the carbon atoms). Another example is the maximum charge of the atoms that are bound to a carbon atom. While for this application it makes sense to describe subsets of atoms by imposing conditions on their element or on the atoms they are bound to, for other applications it is less obvious how to describe useful subsets based on attribute conditions or on related objects. In flow cytometry data analysis for instance, one deals with patient samples (e.g., blood samples), which are described by the individual cells they contain. Each cell in turn is described by some 20 numeric fluorescence intensities. For certain auto-immunity diseases, it is known that the number of cells of a particular cell population (e.g., B-cells, eosinophils,...) is affected. These cell populations are difficult to describe in terms of conditions on the intensity markers, however. They are usually detected by a (manual) clustering procedure over the set of cells. Thus, for such applications, it makes sense to aggregate over subsets defined by clusters instead of defined by attribute conditions.

The contributions of this paper are the following: (1) We present a new type of complex aggregate: aggregates over clusters; (2) We motivate this from a real world example, namely flow cytometry data analysis; (3) We present a first approach to use aggregates over clusters; (4) By showing promising results on both a synthetic and a real world dataset, we motivate future research in this direction.

2   Related Work

Traditionally, relational learners typically handled sets in one of two approaches. The first approach handled sets by checking properties of specific elements. Most inductive logic programming (ILP) [15] systems use this approach, using the existential quantifier to check individual elements. The second approach handled sets by aggregating over them, reducing a complete set to one or a few values (e.g., number of elements in the set, average value of one of the attributes of the set,...). Examples of this approach include probabilistic relational models [13] and relational probability trees [16]. Blockeel and Bruynooghe [5] discussed the resulting - often undesirable - bias on these learners, and proposed the idea of combining aggregation and selection. Challenges in this combination are the large number of features that may be generated and the fact that it is more difficult to traverse the feature space in a structured and efficient way.

Krogel and Wrobel [14] introduced aggregate functions that apply not only to single attributes, but also to pairs of attributes, one of which has to be nominal and serves as a *group by* condition. The resulting aggregate conditions are still of limited complexity and are not refined further during the search. Knobbe et al. [12] proposed a method for subsequently specializing the set to be aggregated. By restricting the application of this specialization operator to aggregate functions where its effect is well-understood, they can search the hypothesis space in a general-to-specific way, but this obviously limits the kind of complex conditions that can be found. Uwents and Blockeel [20] described relational neural networks as a subsymbolic approach towards learning complex aggregates. Their approach is not constrained to using predefined aggregate functions and does not make a distinction between searching for aggregate functions and searching for complex conditions, but the resulting theories are also not interpretable in terms of well-understood aggregates and conditions.

Van Assche et al. [21] have made the first implementation of combined aggregates and selections in an ILP system. They have extended the refinement operator of the relational decision tree learner TILDE [3] to include so-called complex aggregates: literals of the form $F(V, Q, R)$, where F is an aggregate function (e.g., count), V is an aggregate variable occurring in the aggregate query Q, and R is the result of applying F to the set of all answer substitutions for V that Q results in (we will call this set the result set of Q). A complex aggregate is usually followed by a condition on this result set. The "complex" part of a complex aggregate refers to Q being arbitrarily complex. A complex aggregate can be constructed by iterative refinement of Q, starting with a general query (e.g., the number of atoms of a molecule) and ending with a very specific one (e.g., the number of carbon atoms bound with an aromatic bond type to an atom with charge larger than 0.06). The feature explosion resulting from combining aggregate functions with selection conditions was handled by upgrading TILDE to a random forest [21] and taking advantage of the feature sampling applied at each node of the trees. Charnay et al. [6] have recently proposed an alternative solution, by introducing a hill-climbing approach to build complex aggregates incrementally.

Other types of complex aggregates that have been proposed include count-of-count features, which are a kind of nested aggregate, and have been introduced by [9,11] in the type extension tree representation language. An example of such a feature is the complete specification of how many atoms that are present in the molecule to be classified are bound to how many atoms with an aromatic bond type.

Frank et al. [8] have introduced multi-feature aggregation functions in relational learning. In this setting the aggregate variable V is replaced by a set of variables, and this allows the use of aggregate functions like correlation or slope of line of best fit.

3   Case Study: Flow Cytometry Data Analysis

Flow cytometry [1,10] allows to quantify different cell populations in large numbers of cells, by suspending the cells in a stream of fluid and passing them through a laser beam, while measuring the resulting fluorescent and scattered light. Flow cytometry is applied both in clinical settings (in the diagnosis of health disorders) and in research settings. For every input sample, 10,000 up to 1,000,000 cells are measured, each described by 10 up to 20 features. However, recent technological advances in flow cytometry result in techniques that will be able to measure up to 100 dimensions. Standard practice is that pathologists or researchers manually detect different cell populations in the sample by iteratively identifying regions of interest in two-dimensional scatter plots, a process that is labor-intensive and subjective.

One of the first steps in the automatic analysis of this type of data is the use of clustering techniques to identify populations of cells in an objective and reproducible way. Important challenges for clustering these datasets include: populations with different densities; populations that are not elliptic-shaped; rare populations that can be easily confused with noise; hierarchical populations that consist of several sub-types;... Another challenge is that cells can be in transition from one population to another, and thus can belong to several clusters at the same time. As individual patient samples may consist of tens of thousands up to millions of cells, and clinical datasets often consist of hundreds of patient samples, scalability is an important aspect as well. For all these reasons, a number of dedicated clustering techniques for flow cytometry data have been developed [2,7,19,23].

Another important task in the automatic analysis of flow cytometry data is classifying the samples into different groups. This can serve several medical purposes: are they healthy or sick; which variant of the disease do they have; does the medicine work or not? In research settings, the goal often is to learn about which features or cell types are important to differentiate between the groups. Manually, properties of the clusters, such as the relative cell count from a certain cell type, are used to analyze the data. By using automatic data mining techniques, we can make this process easier and more objective. One possible approach is to use propositionalization, based on the cell clusters that have been manually or automatically detected. This requires that the clusters of the different patients are *aligned*, i.e. corresponding clusters need to be identified. This can be done by performing one global clustering, taking the cells of all patients together, and then for each patient indicate the subset of cells belonging to each obtained cluster. However, as cell populations may be shifted from one patient to another, global clustering is likely to perform poorly. Another approach is to perform the clustering individually per patient and then try to match the clusters. However, it is known that certain diseases are characterised by lacking some cell populations, which results in a different number of clusters per patient. These issues have motivated us to explore the use of ILP techniques in this context.

4   Complex Aggregates over Clusters

In this section, we assume a primary table containing the training examples, that has a one-to-many relationship with a secondary table. For a given training example, we want to aggregate over clusters of its related objects in the secondary table. In the flow cytometry use case, the primary table contains the patients, and the secondary table the measured cell properties. Each patient has ten thousands of cells or more in the secondary table. The clusters represent the different cell types present in the patient's sample. When aggregating over clusters instead of over a subset defined by conditions on attributes, there are a number of questions to be addressed.

First, do we search for a similar cluster structure for each example, or do we cluster the objects related to each example independently? The former approach corresponds to propositionalization, and as explained in the previous section, can be performed for instance by clustering over the complete secondary table.

The latter approach requires relational techniques and has the advantage that it is more flexible: the number of clusters can be optimised per example, and it can deal with clusters that are shifted among examples. In this paper we have chosen the second option, and the rest of this discussion also assumes this option.

Second, which type of clustering do we want? One can distinguish total versus partial clustering, overlapping versus non-overlapping clustering, partitional versus hierarchical clustering,... Partitional (flat) clustering algorithms return a flat, unstructured, set of clusters. Hierarchical clustering algorithms, on the other hand, return a cluster hierarchy, the top element corresponding to the complete set of related secondary objects, and the bottom elements to singleton sets.

Third, how do we represent the clustering, and the corresponding complex aggregates? Here, we present two approaches, distinguished by when the clustering step is performed. In general, we would like to express aggregates like "the average value of fluorescent marker M of the cells in a particular cluster is larger than 3", or, in other words, "there exists a cluster of cells with average value for marker M larger than 3". Thus, we obtain the following notation for complex aggregates over clusters: $\exists C, F(V, C, R)$, where C denotes a cluster, and the other variables are as above.

The first approach that we discuss is the case where the clustering step is done in pre-processing. Thus, the clustering is done before and independently of the prediction step. This allows to pre-compute aggregate functions on the obtained clusters. This way, the original data objects of the secondary table can be discarded, retaining only the (aggregated) discovered clusters. In the case of partitional clustering, this highly reduces the cardinality of the relation, resulting in a more efficient prediction step. In the case of hierarchical clustering, this gain depends on the depth of the cluster hierarchy: if the cluster hierarchy is constructed until singleton leaf nodes, the cardinality will be increased instead. One solution is to apply a top-down clustering procedure [4] and use a stopping criterion. A possible disadvantage of this approach is that if there are many examples and many related objects, the clustering step (which is performed for each example independently) may be very time-consuming, while in the final model, only few of the clusterings may actually be used.

The second approach is particularly suited for hierarchical clustering. In this case, a more flexible and efficient refinement strategy is possible, by iteratively performing a clustering step, as illustrated by the following example. Suppose we have a rule stating that a patient has a disease if the sum of the values for marker M of his cells is larger than 50. According to the theory developed in [22], this rule can be specialised in three ways: by decreasing the aggregate function (e.g., replace sum by average or maximum), by increasing the threshold value (e.g., change 50 to 100), or by reducing the set to aggregate on. The latter can be achieved by performing a splitting step on the cells for the patients for which the original rule holds, resulting in a rule that states that there exists a cluster of cells for which the sum of the marker M values is higher than 50. Similarly, a rule stating that there is a cluster of cells with maximum value for marker M smaller than 5 can be specialised a.o. by enlarging the set to aggregate on, i.e., by enlarging the cluster. This can be accomplished by performing a merging step on the clusters for which the original rule holds. This approach leads to a more efficient clustering process than the precomputed (hierarchical) clustering approach above, in the sense that a clustering step (splitting or merging) is only performed "on demand", on the relevant examples, and on the relevant clusters of these examples. On the other hand it requires more memory space than the previous approach, as the original data objects need to be retained.

In this paper we take a first step towards complex aggregates over clusters by exploring the first approach, i.e., using a clustering computed in a pre-processing step.

5   Experiments

In this section, we first describe experiments on two synthetic datasets to show the strengths of aggregating over clusters. Afterwards, we show promising results on a real-world flow cytometry dataset. In all experiments we use a flat clustering approach. We use the relational decision tree learner TILDE[3] to learn the predictive model. We use default parameters for TILDE, except for pruning, which is turned off, and the language bias, which is discussed in detail for each dataset. Thresholds are obtained by discretisation (into ten bins).

5.1   Synthetic Dataset 1

We start the experiments with a simple dataset of 20 samples, each containing 5000 cells and belonging to one of two classes. For illustration purposes, these samples only contain one cluster each. One such dataset is presented in Fig. 1; we generated 6 replicates in total, and report average results.

In many cases, prior background knowledge about the structure of the data is available. In this example, we assume that we know that the clusters can be approximated by ellipses. As such, we fit an ellipse to the data, and use the mean value, the length of the major and minor axis and the angle of the major axis as features to describe the cluster. Note that using the mean value of each dimension, or using the number of cells in the cluster would not discriminate between the two classes.

We have compared four different aggregation levels: checking the existence of individual cells with one of the dimensions smaller or larger than some threshold (*no aggregates*); comparing the maximum, minimum, or average of each

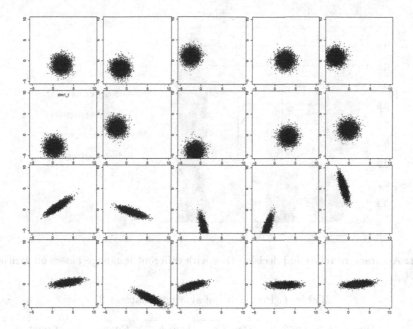

Fig. 1. Synthetic dataset 1.

dimension to a threshold (*simple aggregates*); comparing the count, maximum, minimum, or average of cells with a condition on their dimensions to a threshold (*complex aggregates*); and checking the existence of a cluster with a condition on the number (or percentage) of cells or on the mean, axis lengths or axis angle of the fitted ellipse (*cluster aggregates*).

We compare the different approaches by using them as language bias in the relational decision tree learner TILDE. We use five-fold cross-validation, predicting four patients in each fold. Figure 2 compares accuracy values of the different settings.

We see that the cluster aggregates clearly outperform the other settings. Moreover, the simple and complex aggregates perform worse than selection of individual cells. Inspection of the size of the trees (data not included) shows that the latter yields the largest trees, checking for the existence of cells in different parts of the space. The average number of internal nodes for the different settings in the figure is 3.5, 3.4, 2.6, and 1, respectively. The cluster aggregates are able to classify the data (almost) perfectly with a single test that checks the length of the major or minor axis.

5.2 Synthetic Dataset 2

This dataset reflects the properties of flow cytometry data more closely. It again consists of twenty samples, which might e.g. represent patients, that must be classified into two classes, e.g. healthy or diseased.

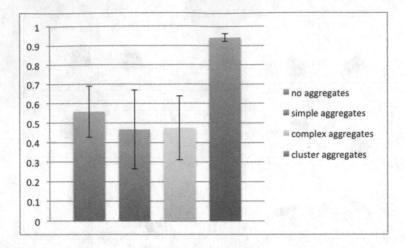

Fig. 2. Accuracy of relational decision tree with different language biases on synthetic dataset 1.

Table 1. Properties of synthetic dataset 2.

	Dim 1		Dim 2		Dim 3		Dim 4		Count	Count
	Mean	Std	Mean	Std	Mean	Std	Mean	Std	class 1	class 2
Cluster 1	2	1.0	2	1.0	4	0.5	2	1.0	5000	5000
Cluster 2	1	0.5	1	0.5	1	0.5	2	1.0	1000 or 2000	1000 or 2000
Cluster 3	1	0.5	3	0.5	1	0.5	2	1.0	1000	1000
Cluster 4	3	0.5	1	0.5	1	0.5	2	1.0	2000 or 1000	2000 or 1000
Cluster 5	3	0.5	3	0.5	1	0.5	2	1.0	**200**	**500**

For each sample, a number of datapoints in a four dimensional space is generated. The datapoints are divided into five clusters, which each have normal distributions. For each patient, the properties of these clusters are slightly different, by perturbing the values in Table 1 using a normal distribution with a small standard deviation around the given values.

As can be seen in the table, cluster five is the important cluster in the diagnosis. Depending on the abundance of datapoints in this cluster, the patients can be split in the two classes. This cluster can be defined as having high values in dimension 1 and 2, low values in dimension 3 and neutral values in dimension 4. Cluster one and four are always present, while clusters two and three can be present in different abundances, without influencing the diagnosis. An illustration of this dataset can be seen in Fig. 3, in total 6 replicates were used, and average results are reported.

To test our algorithmic approach, we again built four different versions of the language bias as input to TILDE. The settings *no aggregates*, *simple aggregates*, and *complex aggregates* are the same as with the previous dataset, except that there are four dimensions now. For the *cluster aggregates*, we assume we have

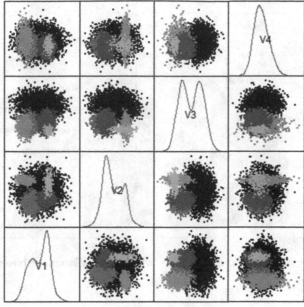

Scatter Plot Matrix

Fig. 3. One sample from synthetic dataset 2. The four-dimensional sample is visualised in all possible two-dimensional plots. The colours indicate the different clusters.

a good clustering algorithm suited for the data. Because we do not want our results to be influenced by the choice of the clustering algorithm, we use the correct clusters, known from generating the data. We describe each cluster by its number of cells and its average value of each dimension.

Accuracy values for five-fold cross-validation can be found in Fig. 4. We see that the simple aggregates and the cluster aggregates perform best. The latter has a large standard deviation, because on one of the replicates, it only obtains an accuracy of 0.3. The complex aggregates yield the smallest trees, but this seems to harm their predictive accuracy.

5.3   HVTN

The HVTN (HIV Vaccine Trials Network) dataset was one of the datasets analysed in the FlowCAP II Challenge [1]. This challenge can be seen as a benchmark tool for sample classification based on flow cytometry data. The study contained 48 individuals, each of them having received an experimental HIV vaccine. Samples were collected approximately 10 months later and T-cells were challenged with two antigens. The response of T-cells was measured by flow cytometry for each of these groups, resulting in two sample files per individuals. The goal of the challenge is to discriminate between the two antigen stimulation groups. Aghaeepour et al. [1] report modest results of previous manual analysis,

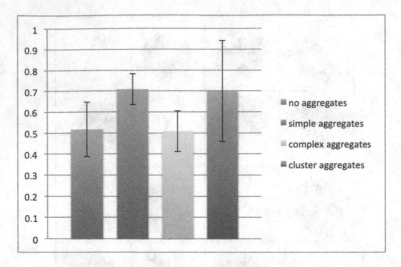

Fig. 4. Accuracy of relational decision tree with different language biases on synthetic dataset 2.

while several classification algorithms that participated in the challenge obtained an (almost) perfect classification, with accuracy ranging from 0.81 to 1.00.

The sample files from 27 individuals were included in the training set, while the files from the other 21 individuals were used as test set. The number of cells that is measured in each file ranges from 40,029 to 354,084.

Experimental Set-Up. We have kept the same train/test separation as in the challenge. On each of the sample files we performed the FLOWMEANS [2] algorithm, which can be seen as an adaptation of k-Means to flow cytometry data. Ten markers (attributes) were used to perform the clustering, the two markers that were left out are known to be unrelated to cell populations. The number of clusters was optimised per sample file, which resulted in 3 to 9 clusters.

For each resulting cluster, the features that we included in TILDE are the absolute and relative cell count, and the mean and standard deviation of each of the ten markers. Each feature was compared to 10 threshold values, obtained by discretisation. Furthermore, we added a complex aggregate that counts the number of clusters with one of these features. Finally, for each of the ten markers, we also added some statistics over the complete sample (i.e., without clustering): mean, standard deviation, skewness, kurtosis, median, and interquartile range. With these features, a bagging ensemble of 100 TILDE trees was run.

An schematic overview of this set-up can be seen in Fig. 5.

Results. Using the procedure described above, we obtained a test set accuracy of 0.93. We are unaware whether the systems that participated in the challenge took into account the fact that for each individual there is one file for each

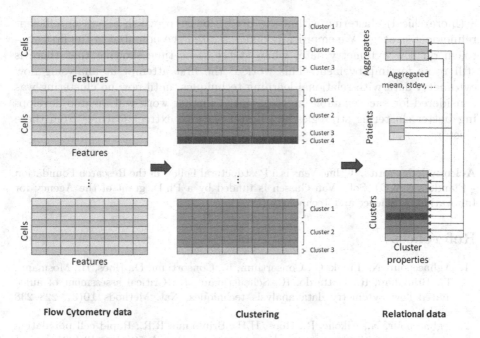

Fig. 5. Schematic overview of the experimental set-up.

antigen stimulation. Since it is known which files belong to which individual, we also evaluated our results using this extra information. For each individual, we checked which prediction was made with the most certainty, and then predicted the opposite class for the other prediction. This resulted in an improved test set accuracy of 0.95. Still, six out of nine systems that participated in the challenge have a better predictive performance, which motivates us to look into better clustering procedures, or better ways to extract features from the predicted clusters, in future work.

We also ran a random forest instead of bagging, using feature sampling at each node of the tree. However, this led to severe overfitting of the training data, which we are also still looking into.

6 Conclusions and Further Work

Flow cytometry data are relationally structured, involving a one-to-many relation with very high cardinality. Handling this relation by selecting specific elements is too fine grained, while aggregating over the set is too coarse grained. This domain clearly needs complex aggregates, where the aggregates are defined over different cell populations. As it is difficult to express such populations in terms of attribute conditions, we here propose to aggregate over clusters. To our knowledge, this has not been considered before. In this paper, we proposed two approaches to aggregate over clusters: either perform a clustering in a pre-processing step, which allows to discard the original data objects, or integrate a (hierarchical) clustering within the prediction step, which allows an efficient refinement strategy. We empirically demonstrated the potential of the first approach on flow cytometry data, while at the same time showing that there is still room for improvement. This work is the first attempt at analysing flow cytometry data with relational learning techniques, until now no clustering was considered for each example independently; further work will include developing better clustering strategies and better ways to extract features from these clusters.

Acknowledgments. Celine Vens is a Postdoctoral Fellow of the Research Foundation - Flanders (FWO). Sofie Van Gassen is funded by a Ph.D. grant of the Agency for Innovation by Science and Technology (IWT).

References

1. Aghaeepour, N., Finak, G., Consortium, F., Consortium, D., Hoos, H., Mosmann, T., Brinkman, R., Gottardo, R., Scheuermann, R.: Critical assessment of automated flow cytometry data analysis techniques. Nat. Methods **10**(3), 228–238 (2013)
2. Aghaeepour, N., Nikolic, R., Hoos, H.H., Brinkman, R.R.: Rapid cell population identification in flow cytometry data. Cytometry Part A **79**(1), 6–13 (2011)
3. Blockeel, H., De Raedt, L.: Top-down induction of first order logical decision trees. Artif. Intell. **101**(1–2), 285–297 (1998)
4. Blockeel, H., De Raedt, L., Ramon, J.: Top-down induction of clustering trees. In: Proceedings of the 15th International Conference on Machine Learning, pp. 55–63 (1998)
5. Blockeel, H., Bruynooghe, M.: Aggregation versus selection bias, and relational neural networks. In: IJCAI-2003 Workshop on Learning Statistical Models from Relational Data, SRL-2003 (2003)
6. Charnay, C., Lachiche, N., Braud, A.: Incremental construction of complex aggregates: Counting over a secondary table. In: Online Preprints of 23th International Conference on Inductive Logic Programming, pp. 1–6 (2013)
7. Finak, G., Bashashati, A., Brinkman, R., Gottardo, R.: Merging mixture components for cell population identification in flow cytometry. Adv. Bioinform. **2009**, 12 (2009)
8. Frank, R., Moser, F., Ester, M.: A method for multi-relational classification using single and multi-feature aggregation functions. In: Kok, J.N., Koronacki, J., Lopez de Mantaras, R., Matwin, S., Mladenič, D., Skowron, A. (eds.) PKDD 2007. LNCS (LNAI), vol. 4702, pp. 430–437. Springer, Heidelberg (2007)
9. Frasconi, P., Jaeger, M., Passerini, A.: Feature discovery with type extension trees. In: Železný, F., Lavrač, N. (eds.) ILP 2008. LNCS (LNAI), vol. 5194, pp. 122–139. Springer, Heidelberg (2008)
10. Herzenberg, L., Tung, J., Moore, W., Herzenberg, L., Parks, D.: Interpreting flow cytometry data: a guide for the perplexed. Nat. Immunol. **7**(7), 681–685 (2006)
11. Jaeger, M., Lippi, M., Passerini, A., Frasconi, P.: Type extension trees for feature construction and learning in relational domains. Artif. Intell. **204**, 30–55 (2013)
12. Knobbe, A.J., Siebes, A., Marseille, B.: Involving aggregate functions in multi-relational search. In: Elomaa, T., Mannila, H., Toivonen, H. (eds.) PKDD 2002. LNCS (LNAI), vol. 2431, p. 287. Springer, Heidelberg (2002)

13. Koller, D.: Probabilistic relational models. In: Džeroski, S., Flach, P.A. (eds.) ILP 1999. LNCS (LNAI), vol. 1634, p. 3. Springer, Heidelberg (1999)
14. Krogel, M.A., Wrobel, S.: Facets of aggregation approaches to propositionalization. In: Horváth, T., Yamamoto, A. (eds.) Proceedings of the Work-in-Progress Track at the 13th International Conference on Inductive Logic Programming, pp. 30–39 (2003)
15. Muggleton, S. (ed.): Inductive Logic Programming. Academic Press, New York (1992)
16. Neville, J., Jensen, D., Friedland, L., Hay, M.: Learning relational probability trees. In: Proceedings of the 9th ACM SIGKDD International Conference on Knowledge Discovery and Data Mining, pp. 625–630. ACM Press (2003)
17. Perlich, C., Provost, F.: Aggregation-based feature invention and relational concept classes. In: Proceedings of the 9th ACM SIGKDD International Conference on Knowledge Discovery and Data Mining, pp. 167–176. ACM Press (2003)
18. Srinivasan, A., Muggleton, S., King, R.: Comparing the use of background knowledge by inductive logic programming systems. In: De Raedt, L. (ed.) Proceedings of the 5th International Workshop on Inductive Logic Programming, pp. 199–230 (1995)
19. Sugár, I.P., Sealfon, S.C.: Misty mountain clustering: application to fast unsupervised flow cytometry gating. BMC Bioinf. 11(1), 502 (2010)
20. Uwents, W., Blockeel, H.: Classifying relational data with neural networks. In: Kramer, S., Pfahringer, B. (eds.) ILP 2005. LNCS (LNAI), vol. 3625, pp. 384–396. Springer, Heidelberg (2005)
21. Van Assche, A., Vens, C., Blockeel, H., Džeroski, S.: First order random forests: learning relational classifiers with complex aggregates. Mach. Learn. 64(1–3), 149–182 (2006)
22. Vens, C., Ramon, J., Blockeel, H.: Refining aggregate conditions in relational learning. In: Fürnkranz, J., Scheffer, T., Spiliopoulou, M. (eds.) PKDD 2006. LNCS (LNAI), vol. 4213, pp. 383–394. Springer, Heidelberg (2006)
23. Zare, H., Shooshtari, P., Gupta, A., Brinkman, R.R.: Data reduction for spectral clustering to analyze high throughput flow cytometry data. BMC Bioinf. 11(1), 403 (2010)

On the Complexity of Frequent Subtree Mining in Very Simple Structures

Pascal Welke[1]([✉]), Tamás Horváth[1,2], and Stefan Wrobel[1,2]

[1] Department of Computer Science, University of Bonn, Bonn, Germany
welke@uni-bonn.de
[2] Fraunhofer IAIS, Schloss Birlinghoven, Sankt Augustin, Germany

Abstract. We study the complexity of frequent subtree mining in very simple graphs beyond forests. We show for d-tenuous outerplanar graphs that frequent subtrees can be listed with polynomial delay if the *cycle degree*, i.e., the maximum number of blocks that share a common vertex, is bounded by some *constant*. The crucial step in the proof of this positive result is a polynomial time algorithm deciding subgraph isomorphism from trees into d-tenuous outerplanar graphs of *bounded* cycle degree. We obtain this algorithm by generalizing the algorithm of Shamir and Tsur that decides subgraph isomorphism between trees. Our results may also be of some interest to algorithmic graph theory, as they indicate that even for very simple structures, the cycle degree is a crucial parameter for the tractability of subgraph isomorphism. We also discuss some interesting problems towards generalizing the positive result of this work.

1 Introduction

One of the central problems of graph mining is the *frequent connected subgraph mining* (FCSM) problem defined as follows: *Given* a set D of graphs, called *transaction graphs*, from some graph class $\mathcal{G}$ and some integral frequency threshold $t > 0$, *generate* all connected graphs that are subgraph isomorphic to at least t graphs in D. We require the patterns in the output to be unique up to isomorphism.

This problem is of high relevance to Inductive Logic Programming. Indeed, each transaction graph $G \in D$ can be regarded as an intensional database D_G over a relational vocabulary consisting of a set of constants, a set of unary, and a set of binary predicate symbols corresponding to the vertices, the vertex labels, and the edge labels of the graphs in D, respectively. Furthermore, all patterns P can be viewed as goal clauses, or equivalently, as Boolean conjunctive queries over the same vocabulary. Since for any such P and D_G, the first-order logical implication $P \models D_G$ becomes equivalent to subsumption between the clauses corresponding to P and D_G [7], which, in turn, is equivalent to relational homomorphism (cf. [14]), the FCSM problem can be regarded as a special case of the problem of generating all frequent Boolean conjunctive queries in the learning from interpretations setting with the additional constraint that the

© Springer International Publishing Switzerland 2015
J. Davis and J. Ramon (Eds.): ILP 2014, LNAI 9046, pp. 194–209, 2015.
DOI: 10.1007/978-3-319-23708-4_14

logical consequence (or pattern matching) operator is defined by subsumption under object identity (see, e.g., [4, 6]).

The FCSM problem has been intensively studied over the past two decades by the data mining community, with the primary goal of designing practical algorithms for different problem settings. Despite the intensive research effort, its complexity aspects are still not well understood. For example, there are only a few non-trivial positive results about special cases of the FCSM problem when frequent connected subgraphs are required to be generated with polynomial delay. Polynomial delay pattern enumeration means that the delay between generating two consecutive patterns is bounded by a polynomial in the size of D. One of the well-known such cases is that the transaction graphs are restricted to forests [3]. The goal of this work is to generalize this positive result. In particular, we concentrate on the following question that, to the best of our knowledge, has not been investigated so far: *How far can we go beyond forests when insisting on polynomial delay frequent pattern generation?* We study this question by looking at very simple transaction and pattern classes.

In particular, we focus on a subclass of *d-tenuous outerplanar* graphs. A graph is *outerplanar* if it can be drawn in the plane in a such way that edges are allowed to intersect each other only in their endpoints and all vertices can be reached from outside without crossing any edge. An outerplanar graph is *d-tenuous* if it has at most d diagonals per block. This later definition uses the fact that all blocks (i.e., biconnected components with more than one edge) of an outerplanar graph can be drawn in the plane as a convex polygon containing some non-crossing diagonals. We note that d-tenuous outerplanar graphs form a practically relevant graph class. For example, the molecular graphs of most pharmacological compounds are d-tenuous outerplanar graphs for some small d [13].

The main result of this paper is that frequent subtrees can be listed with polynomial delay from d-tenuous outerplanar graphs of bounded *cycle degree*. The cycle degree of a graph is the maximum number of blocks a vertex can belong to. Our mining algorithm is an adaptation of a generic levelwise search algorithm to d-tenuous outerplanar graphs of bounded cycle degree. The most complex part of the algorithm is the pattern matching subroutine. It utilizes that any block of a d-tenuous outerplanar graph has polynomially many spanning trees only. This property, combined with bounded cycle degree, allows for an efficient algorithm deciding subgraph isomorphism. Though a d-tenuous outerplanar graph of bounded cycle degree may have exponentially many spanning trees, we show that any instance of our more general problem can be decomposed with a divide-and-conquer strategy into polynomially many instances of deciding subgraph isomorphism between trees, which can be solved in polynomial time [18]. Our technique requires an efficient combination of the blocks' spanning trees, as well as a careful assembly of certain partial subgraph isomorphisms.

To put our result in context, we first note that frequent connected subgraph mining is possible with polynomial delay if the pattern matching is restricted to a constrained subgraph isomorphism that maps bridges to bridges and different blocks to different blocks [13]. This directly implies that frequent subtrees can be listed with polynomial delay if D consists of forests (see, also, [3]). For bounded tree-width graphs, a proper generalization of forests, frequent subtrees can be generated in incremental polynomial time [12] for ordinary subgraph isomorphism. That is, a next new frequent subtree for this graph class can be computed in time polynomial in the combined size of the input and the set of patterns already computed. Beyond forests, however, it is an open question whether polynomial delay mining of frequent subtrees is possible even for graphs of bounded tree-width, if the vertex degree is not bounded by some constant. In fact, the question remains open even for the very simple case of 0-tenuous outerplanar graphs, also known as *cactus* graphs. An affirmative answer to this question would immediately solve the open problem whether polynomial delay pattern generation is possible for NP-hard pattern matching operators, as it is an NP-complete problem to decide whether a tree is subgraph isomorphic to a cactus graph [2]. On the other hand, as shown in Sect. 4.1, a negative answer would imply that P $\neq$ NP. Regarding the two cases above, we conjecture that polynomial delay frequent pattern generation is *not* possible for computationally intractable pattern matching operators; this is certainly the case for graph classes for which the Hamiltonian path problem is NP-complete. If this conjecture holds then the results of this paper imply that *even for very simple structures, the cycle degree of graphs is a crucial parameter* that must be taken into account when designing efficient frequent subgraph mining algorithms. Furthermore, polynomial delay generation of frequent acyclic Boolean conjunctive queries (i.e., Datalog goal clauses) is possible only for very simple relational structures in the learning from interpretation setting with subsumption under object identity [4,6].

The rest of the paper is organized as follows. In Sect. 2 we collect the necessary background and then describe our mining algorithm in Sect. 3. In Sect. 4 we conclude and discuss some interesting problems towards generalizing our results. Due to space limitations we omit the proof of the correctness and only sketch the runtime of our pattern matching algorithm.

2   Notions

In this section we fix the terminology and notation used in the paper. We first recall some standard notions from graph theory (see, e.g., [5]) and then, in Sect. 2.1, formally define the mining problem considered in this work.

An *undirected* (resp. *directed*) *graph* G consists of a finite set $V(G)$ of vertices and a set $E(G) \subseteq \{X \subseteq V(G) : |X| = 2\}$ (resp. $E(G) \subseteq V(G) \times V(G)$) of edges. We consider simple graphs, i.e., loops and parallel edges are not permitted. Unless otherwise stated, by *graphs* we mean undirected graphs and denote an edge $\{u, v\} \in E(G)$ by uv. The set of neighbors of a vertex v is denoted by $\mathcal{N}(v)$. A *subgraph* of G is a graph G' with $V(G') \subseteq V(G)$ and $E(G') \subseteq E(G)$; G' is a subgraph of G *induced* by a set $V' \subseteq V(G)$ if $V(G') = V'$ and $uv \in E(G')$ if and only if $uv \in E(G)$ for all $u, v \in V'$. Such an induced subgraph is denoted by $G[V']$. A *labeled* graph is a graph G such that all vertices and all edges are labeled with some symbols from a finite set. Examples of labeled graphs include molecular graphs, protein-protein interaction graphs, the Web graph etc. To keep the notation and description concise, we will state all our results for unlabeled graphs by noting that all our arguments apply to labeled graphs as well.

Two graphs G_1, G_2 are *isomorphic*, if there is a bijection $\varphi : V(G_1) \to V(G_2)$ such that $uv \in E(G_1)$ iff $\varphi(u)\varphi(v) \in E(G_2)$ for all $u, v \in V(G_1)$. G_1 is *subgraph isomorphic* to G_2, denoted $G_1 \preccurlyeq G_2$, if G_2 has a subgraph isomorphic to G_1.

A *graph class* is a set of pairwise non-isomorphic graphs that share some common property, e.g., they have tree-width at most k for some integer $k > 0$. In this paper we will concentrate on the class of outerplanar graphs; a graph is *outerplanar*, if (i) it can be drawn in the plane in such a way that edges do not cross each other except maybe in their endpoints and (ii) every vertex lies on the outer face. That is, each vertex can be reached from outside without crossing any edge. A graph G is outerplanar if and only if all its biconnected components are outerplanar [10]. A *biconnected component* B of an outerplanar graph is either a single edge not belonging to a cycle, called *bridge*, or it is a maximal induced subgraph of G composed of a single Hamiltonian cycle and possibly some diagonal edges [10]. Biconnected components of the later type will be referred to as *blocks*. A *d-tenuous* outerplanar graph is an outerplanar graph in which each block has at most d diagonals. Notice that forests are special outerplanar graphs.

2.1   The Frequent Connected Subgraph Mining Problem

In this work, we study special cases of the following problem:

Frequent Connected Subgraph Mining (FCSM) Problem: *Given* a finite set $D \subseteq \mathcal{G}$ for some graph class $\mathcal{G}$ and an integer threshold $t > 0$, *list* all graphs $P \in \mathcal{P}$ for some graph class $\mathcal{P}$, called the pattern class, that are subgraph isomorphic to at least t graphs in D.

The patterns in the output must be pairwise non-isomorphic. In contrast to the standard problem definition (see, e.g. [12]), we regard a more general problem *parameterized* by the pattern class. Though in this paper we are interested in the case that $\mathcal{P}$ is the class of trees, we use here this more general problem setting. The reason is that in Sect. 3.1 we give an algorithm for this generic problem and state conditions guaranteeing efficient pattern mining. These conditions may be of some independent interest.

Note that the problem above is a listing problem. For such problems, the following complexity classes are distinguished in the literature (see, e.g., [15]). Suppose an algorithm $\mathfrak{A}$ for the FCSM problem gets D and t as input and outputs a list $O = [p_1, p_2, \cdots, p_n]$ of patterns. Then $\mathfrak{A}$ generates O

– with *polynomial delay*, if the time before the output of p_1, between the output of consecutive elements of O, and between the output of p_n and the termination of $\mathfrak{A}$ is bounded by a polynomial of $size(D)$,

– in *incremental polynomial time*, if the algorithm outputs p_1 in time bounded by a polynomial of $size(D)$, the time between outputting p_i and p_{i+1} is bounded by a polynomial of $size(D) + \sum_{j=1}^{i} size(p_j)$, and the time between the output of p_n and termination is bounded by a polynomial of $size(D) + size(O)$.

Clearly, polynomial delay implies incremental polynomial time. It is an open problem whether the two classes are identical, or not. In frequent itemset mining, for example, the FP-Growth algorithm [9] lists frequent patterns with polynomial delay, while the Apriori algorithm [1] in incremental polynomial time. Our special focus in this work is to establish sufficient conditions for polynomial delay pattern mining.

3   The Mining Algorithm

In this section we present our algorithm mining frequent tree patterns in d-tenuous outerplanar graphs of bounded cycle degree with polynomial delay. We denote the transaction graph class to be considered by $\mathcal{O}_{d,c}$, where c is an upper bound on the cycle degree of the graphs it contains. More precisely, let G be a graph and $\mathcal{B}(G)$ the set of its blocks. Then the number of distinct blocks containing a vertex $v \in V(G)$, denoted by $\delta_C(v)$, is called the *cycle degree* of v. That is, $\delta_C(v) := |\{B \in \mathcal{B}(G) : v \in V(B)\}|$. We define the cycle degree of a graph to be the maximum cycle degree over all its vertices. Notice that the *vertex degree* of the graphs in $\mathcal{O}_{d,c}$ can be unbounded.

The class $\mathcal{O}_{d,c}$ is not only of theoretic, but also of practical interest e.g. in computational chemistry. For example, for the NCI2012 dataset[1] consisting of $265,242$ molecular graphs we have the following cycle degree distribution:

cycle degree	#graphs
0	21,727
1	238,218
2	5,099
3	196
4	2

Additionally, most molecular graphs have an outerplanar graph structure with a small number of diagonals per block [13].

3.1   A Generic Levelwise Search Mining Algorithm

We obtain our positive result by adapting a generic levelwise search mining algorithm to our problem setting. Levelwise search [1] is one of the most popular techniques in pattern mining that can be used to efficiently mine frequent patterns in a broad range of problem settings. In order to find a pattern in level $l+1$,

---

[1] http://cactus.nci.nih.gov/download/nci, Release 4 File Series.

it completely explores all levels up to l. On the one hand, this is disadvantageous if one is interested in frequent patterns only from level l for some large l, on the other hand, in frequent subgraph mining it allows for an incremental polynomial time pattern generation even for NP-complete pattern matching operators [12].

Input: $D \subseteq \mathcal{G}$ for some graph class $\mathcal{G}$, a pattern class $\mathcal{P}$, and an integer $t > 0$
Output: all frequent subgraphs of D that are in $\mathcal{P}$

```
1: let S₀ ⊆ P be the set of frequent pattern graphs consisting of a single vertex
2: for (l := 0; Sₗ ≠ ∅; l := l + 1) do
3:     set Sₗ₊₁ := ∅ and Cₗ₊₁ := ∅
4:     for all P ∈ Sₗ do
5:         print P
6:         for all H ∈ ρ(P) ∩ P satisfying H ∉ Cₗ₊₁ do
7:             add H to Cₗ₊₁
8:             if SUPPORTCOUNT(H, D) ≥ t then
9:                 add H to Sₗ₊₁
```

Algorithm 1: A generic levelwise graph mining algorithm.

Algorithm 1 is a generic levelwise search algorithm for the FCSM problem. It is a slight modification of a related algorithm in [12]; the only changes are in Lines 1 and 6. It calculates the set of candidate (resp. frequent) patterns of level l in the set variable C_l (resp. S_l). In Line 6 it computes the set $\rho(P)$ of refinements of a pattern P obtained from P by extending it with an edge in all possible ways. That is, it either adds a new vertex w to P and connects it to any vertex in $V(P)$ by an edge, or it connects two vertices in $V(P)$ that have not been connected yet. Clearly, $|\rho(P)| \in O\left(|V(P)|^2\right)$. Subroutine SUPPORTCOUNT(H, D) in Line 8 returns the number of graphs $G \in D$ with $H \preccurlyeq G$.

It is shown in [12] that the original version of Algorithm 1 mines frequent patterns with polynomial delay if patterns and transactions satisfy certain conditions. These conditions have however been formulated for the case that the pattern and transaction graph classes are the same. In the theorem below we generalize these conditions to the case that they can be different. Due to space limitation, we only sketch the proof.

Theorem 1. *Let $\mathcal{G}$ and $\mathcal{P}$ be the transaction and pattern graph classes satisfying the following conditions:*

1. *All graphs in $\mathcal{P}$ are connected. Furthermore, $\mathcal{P}$ is closed downwards under taking subgraphs, i.e., for all $H \in \mathcal{P}$ and for all connected graphs H' we have $H' \in \mathcal{P}$ whenever $H' \preccurlyeq H$.*
2. *The membership problem for $\mathcal{P}$ can be decided efficiently, i.e., for any graph H it can be decided in polynomial time if $H \in \mathcal{P}$.*
3. *Subgraph isomorphism in $\mathcal{P}$ can be decided efficiently, i.e., for all $H_1, H_2 \in \mathcal{P}$, it can be decided in polynomial time if $H_1 \preccurlyeq H_2$.*
4. *Subgraph isomorphism between patterns and transactions can be decided efficiently, i.e., for all $H \in \mathcal{P}$ and $G \in \mathcal{G}$, it can be decided in polynomial time if $H \preccurlyeq G$.*

Then the FCSM problem can be solved with polynomial delay for $\mathcal{P}$ and for all finite subsets $D \subseteq \mathcal{G}$.

Proof. Let $\mathcal{G}$ and $\mathcal{P}$ be two graph classes such that Conditions 1–4 hold. Let P be a graph. If $P \notin \mathcal{P}$ or P is not frequent in D then it cannot be part of the output of Algorithm 1 (see Lines 6 and 8, respectively). Now let $P \in \mathcal{P}$ be frequent in D. If P has more than one vertex, there is a vertex with degree one or there is an edge that can be removed without disconnecting P. Inductively, we get a sequence of edge (and isolated vertex) removals that in reverse order is a sequence of refinements generating P from a single vertex. By Condition 1, every connected subgraph of P is in $\mathcal{P}$ and is also frequent. Thus, by induction, Algorithm 1 generates and outputs P. The condition in Line 6 ensures that the algorithm outputs any pattern P at most once up to isomorphism and that P will not be processed if $P \notin \mathcal{P}$. Thus the LEVELWISEGRAPHMINING algorithm is correct.

Regarding its runtime, the time between starting and outputting the first pattern (or termination if there is no frequent single vertex) is linear in the size of the database. We can count the frequency of single vertices by a single scan over the database. We now show that the time needed for Lines 6–9 is polynomial in the size of D. Conditions 1–3 imply that there is a canonical string representation for all graphs in $\mathcal{P}$ that can be computed in polynomial time. We can store $\mathcal{S}_i$ and $\mathcal{C}_i$ as prefix trees of canonical strings of patterns. In this way, we can add and look up patterns in $\mathcal{S}_i$ or $\mathcal{C}_i$ in time linear in the size of the canonical string of a pattern. $|\rho(P)|$ is polynomial in the size of P and thus, in the size of D. Therefore, by Condition 2, $\rho(P) \cap \mathcal{P}$ can be computed in polynomial time. $H \notin \mathcal{C}_{l+1}$ can be checked in time linear in the size of the canonical string representation of H. SUPPORTCOUNT can be implemented by iterating over D, checking for each graph $G \in D$ if $H \preccurlyeq G$, and maintaining a counter; by Condition 4 it runs in polynomial time. Overall, the time between printing consecutive patterns and the time between printing the last pattern and termination is polynomial in the size of D. $\square$

Notice that the conditions above allow for generating frequent patterns that do not belong to $\mathcal{G}$. Furthermore, they enable the generation of restricted subsets of *all* frequent subgraphs. For example, we can mine frequent paths in transaction databases consisting of trees. We will utilize the later property when restricting $\mathcal{P}$ to trees. For this case, one can check that $\mathcal{P}$ satisfies the first three conditions (Conditions 1 and 2 are straightforward for trees and Condition 3 follows e.g. from [18]). Thus, to show that tree patterns can be mined with polynomial delay in any database of graphs from some graph class $\mathcal{G}$, we only need to show that Condition 4 holds for trees and for the graphs in $\mathcal{G}$. We state this for $\mathcal{G} = \mathcal{O}_{d,c}$ in Theorem 3 in Sect. 3.2, immediately implying our main result for this work:

Theorem 2. *All frequent subtrees can be generated with polynomial delay in transaction databases from $\mathcal{O}_{d,c}$ for constant d and c.*

3.2   The Algorithm Deciding Subgraph Isomorphism

The main result of this section is formulated in Theorem 3. Due to space limitations, we omit all proofs and describe only the algorithm implying Theorem 3.

Theorem 3. *For any tree H and $G \in \mathcal{O}_{d,c}$, it can be decided in polynomial time whether $H \preccurlyeq G$.*

For the rest of this section, let $G \in \mathcal{O}_{d,c}$ be a graph with n vertices and H be a tree with $k > 1$ vertices. We can assume without loss of generality that $k \leq n$, as there is no subgraph isomorphism from H to G whenever $|V(H)| > |V(G)|$, and that G is connected, as any subgraph isomorphism maps a connected graph into a connected component. We fix an arbitrary vertex $r \in V(G)$ and will denote the choice of r by G^r. For a biconnected component B we define its *root*, denoted $\rho_r(B)$, to be the vertex of B with the smallest distance to r, where by distance between two vertices we mean the length of a shortest path connecting them. Since B is a maximal biconnected subgraph, $\rho_r(B)$ is unique. Furthermore, r is always among the roots of biconnected components (recall that bridges are also biconnected components). B will also be referred to as a *v-rooted* biconnected component, where $v = \rho_r(B)$. The subgraph formed by the set of v-rooted biconnected components of G^r is denoted by $\mathcal{B}_v^r$ (we omit G from the notation as it will always be clear from the context). On the set of roots of all biconnected components in G^r we define a directed graph T_G^r as follows: For any $u, v \in V(T_G^r)$ with $u \neq v$ we have $(u, v) \in E(T_G^r)$ if and only if there exists a biconnected component B in $\mathcal{B}_v^r$ with $u \in V(B)$. T_G^r is a tree (rooted at r) that will be traversed by the algorithm deciding subgraph isomorphism. Figure 1 (a) and (b) show an example of the structures G^r and T_G^r for a small graph G.

A vertex w is *below* v with respect to r if all paths in G from r to w contain v. If r is clear from the context we simply say w is below v. The definitions imply that every vertex is below itself and that all vertices are below r. A *rooted subgraph* G_v^r of G for a vertex $v \in V(G)$ is the subgraph of G induced by the set of vertices below v, i.e.

$$G_v^r = G\left[\{w : w \text{ is below } v\}\right].$$

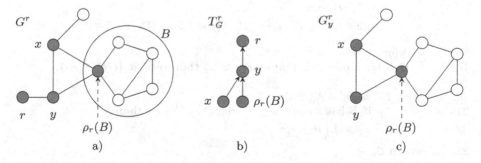

a)   b)   c)

Fig. 1. G^r, T_G^r and G_y^r for a small graph G. $\rho_r(B)$ is the root of the block B. Roots are shown in gray, while vertices that are not roots are shown in white.

See Fig. 1 (c) for an example of G_v^r. It follows from the definitions and from the connectivity of G that G_v^r is a connected graph. Furthermore, the definitions above imply that $G_r^r = G$ and that G_w^r is a single vertex if and only if $w \notin V(T_G^r)$.[2] A vertex $w \in V(G)$ is called a *child* of v, if $vw \in E(G)$ and w belongs to a v-rooted biconnected component.

Using the above notions, we now describe our subgraph isomorphism algorithm given in Algorithm 2. It utilizes the fact that for any free tree H and connected graph G, $H \preccurlyeq G$ if and only if H is subgraph isomorphic to a spanning tree of G. Since G can have exponentially many spanning trees, we cannot directly apply the subtree isomorphism algorithm given in [18] to *all* spanning trees of G. For every vertex v, however, the graph induced by all biconnected components containing v has only a polynomial number of spanning trees in the size of G, since both d and c are bounded by constants. Our algorithm decides subgraph isomorphism by computing and combining (partial) subgraph isomorphisms from subtrees of the pattern to such spanning trees rooted at v in a bottom up way.

1: **function** MAIN(H, G)
2:     pick a vertex $r \in V(G)$ and set $\mathcal{C} := \emptyset$
3:     **for all** roots v of T_G^r in a postorder **do**
4:         let $\mathcal{S}_v^r$ be the set of spanning trees of $\mathcal{B}_v^r$, each rooted at v
5:         **for all** $\tau \in \mathcal{S}_v^r$ **do**
6:             **for all** $w \in V(\tau)$ in a postorder **do**
7:                 **for all** $u \in V(H)$ **do**
8:                     **if** $w \in V(\tau) \setminus \{v\}$ or $w = r$ **then**
9:                         $\mathcal{C} := \mathcal{C} \cup$ CHARACTERISTICS(u, v, τ, w)
10:                        **if** $(H_u^u, \tau, w) \in \mathcal{C}$ **then return** TRUE
11:    **return** FALSE

12: **function** CHARACTERISTICS(u, v, τ, w)
13:     $\mathcal{C}_\tau := \emptyset$
14:     **for all** $\theta \in \Theta_{vw}(\tau)$ **do**
15:         let τ' be the tree satisfying $\theta = \tau \cup \tau'$
16:         let C_τ (resp. $C_{\tau'}$) be the children of w in τ (resp. τ') and $C_\theta := C_\tau \cup C_{\tau'}$
17:         let $B = (C_\theta \dot\cup \mathcal{N}(u), E)$ be the bipartite graph with

$$cu' \in E \iff$$
$$(c \in C_\tau \wedge (H_{u'}^u, \tau, c) \in \mathcal{C}) \vee (c \in C_{\tau'} \wedge (H_{u'}^u, \tau', c) \in \mathcal{C})$$

        for all $cu' \in C_\theta \times \mathcal{N}(u)$
18:         **if** B has a matching that covers $\mathcal{N}(u)$ **then return** $\{(H_u^u, \tau, w)\}$
19:         **else**
20:             **for all** $u' \in \mathcal{N}(u)$ **do**
21:                 **if** B has a matching covering $\mathcal{N}(u) \setminus \{u'\}$ **then**
22:                     add $\left(H_u^{u'}, \tau, w\right)$ to $\mathcal{C}_\tau$
23:     **return** $\mathcal{C}_\tau$

Algorithm 2: The Subtree Isomorphism Algorithm

---

[2] We recall that by condition, G contains at least one edge.

We now formally define the type of partial subgraph isomorphisms used by our algorithm.[3] Let G' be an induced subgraph of G and τ be a spanning tree of G'. Then $G|_\tau$ denotes the graph obtained from G by removing all edges of G' that are not in τ. We will often choose $G' = \mathcal{B}_v^r$ for some root v and spanning tree τ of $\mathcal{B}_v^r$. For this case $G_v^r|_\tau$ denotes the graph obtained from G_v^r by removing all edges from $\mathcal{B}_v^r$ that are not in τ. As an example, $G_v^r|_\tau$ in Fig. 2 arises from G by removing the edges rv, vx and the vertex r. Finally, an *iso-triple* ξ of H relative to a root $v \in V(T_G^r)$ is a triple (H_u^y, τ, w), where $u \in V(H)$, $y \in \mathcal{N}(u) \cup \{u\}$, τ is a spanning tree of $\mathcal{B}_v^r$, and $w \in V(\tau)$.

Definition 1. *An iso-triple ξ is a v-characteristic if and only if there exists a subgraph isomorphism φ from H_u^y to $(G_v^r|_\tau)_w^v$ with $\varphi(u) = w$.*

In Lemma 1 below we first provide a characterization of subgraph isomorphisms from trees into outerplanar graphs in terms of v-characteristics. Its proof follows directly from the definitions.

Lemma 1. *Let H be a tree, G be an outerplanar graph, and r be an arbitrary vertex of G. Then $H \preccurlyeq G$ if and only if there exists a v-characteristic (H_u^u, τ, w) for some $v \in T_G^r$, $u \in V(H)$, spanning tree τ of $\mathcal{B}_v^r$, and $w \in V(\tau)$.*

As an example, consider the star graph H with center u and neighbors u_1, u_2, u_3, u_4, u_5 and the graph G^r in Fig. 2. Obviously, there is a subgraph isomorphism from H to G mapping u to w. For each leaf u_i of H ($u_i = u_1$ in Fig. 2), we would find a v-characteristic $(H_u^{u_i}, \tau, w)$, as we can injectively map the four remaining neighbors of u to the children of w in τ and τ', i.e. to the neighbors of w in G, except for v. When we compute the r-characteristics for v later on, we will then find a characteristic $(H_{u'}^{u'}, \tau'', v)$ for each leaf u' of H and unique spanning tree τ'' of B_r^r with the single edge vr. Note that it is crucial in this example to combine the spanning trees of $\mathcal{B}_v^r$ with those of $\mathcal{B}_w^r$ or $\mathcal{B}_r^r$, as we would otherwise not be able to find a subgraph isomorphism from H to G.

Our algorithm, depicted in Algorithm 2, utilizes the condition stated in Lemma 1 above. It selects an arbitrary vertex $r \in V(G)$ (Line 2) and calculates the characteristics for all root vertices by traversing T_G^r in a postorder manner (Line 3). For any $v \in V(T_G^r)$, the set of v-characteristics can be computed, as we discuss below, from the sets of characteristics of the children of v in T_G^r, by considering all spanning trees of $\mathcal{B}_v^r$. Since G is a d-tenuous outerplanar graph of bounded cycle degree, the number of spanning trees of $\mathcal{B}_v^r$ is bounded by a polynomial of the size of G. Thus, for all $v \in V(T_G^r)$ visited, Algorithm 2 first computes the set $\mathcal{B}_v^r$ of v-rooted biconnected components and then calculates the set $\mathcal{S}_v^r$ of all spanning trees of $\mathcal{B}_v^r$ (Line 4). Each spanning tree is regarded as a tree rooted at v. The root v itself, if it is different from r, is ignored at the stage when the algorithm visits it, as it will be processed when visiting its

---

[3] Though the terminology below follows that of [8] (which, in turn, is based on the concepts in [16]), the definitions of iso-triples and characteristics in this paper are entirely different from those in [8].

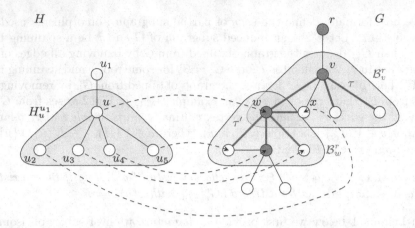

Fig. 2. This figure shows a small graph G^r with its subgraphs $\mathcal{B}_v^r$ and $\mathcal{B}_w^r$ (depicted by the rounded triangles) and the star graph H. One spanning tree τ of $\mathcal{B}_v^r$ and τ' of $\mathcal{B}_w^r$ are shown in red and blue, respectively (Color figure online).

parent vertex in T_G^r (Lines 7 and 8). Otherwise, i.e., when $v = r$, we need to compute the r-characteristics at the end of the postorder traversal of τ, i.e., when $r = v = w$, as r has no parent root.

In Loop 3–10 we calculate in variable $\mathcal{C}$ the set of v-characteristics for all combinations of $v \in V(T_G^r)$, spanning tree τ of $\mathcal{B}_v^r$, $u \in V(H)$, and $w \in V(\tau)$. By Lemma 1, there exists a subgraph isomorphism from H to G if and only if there are v, τ, u, and w such that the iso-triple (H_u^u, τ, w) is a v-characteristic. This condition is tested in Line 10. If such a v-characteristic exists the algorithm terminates by returning the answer TRUE. Thus, the algorithm reaches Line 11 and returns FALSE only if there is no subgraph isomorphism from H to G.

We now turn to the problem of computing the characteristics for the nodes of T_G^r. In Lemmas 2 and 3 below we formulate sufficient and necessary conditions for iso-triples being characteristics. These conditions, as we show below, allow for a recursive computation of the set of v-characteristics for any node $v \in V(T_G^r)$ from those of its children in T_G^r and with respect to the spanning trees of $\mathcal{B}_v^r$, justifying the post-order traversal of T_G^r (see Line 3 of Algorithm 2). To compute the v-characteristics for a root vertex v with respect to a spanning tree τ of $\mathcal{B}_v^r$, our algorithm traverses τ in a post-order manner and checks whether there exists a subgraph isomorphism from a certain subtree of H to $G_v^r|_\tau$, i.e., to the graph obtained from G_v^r by replacing $\mathcal{B}_v^r$ with its spanning tree τ. Since $V(\tau) = V(\mathcal{B}_v^r)$, this operation is invariant to the set of vertices of G_v^r. The difficulty of this step is, however, that $\mathcal{B}_v^r$ does not contain the complete neighborhood of w whenever $w \in V(\tau)$ with $w \neq r$ is a root vertex of another biconnected block in G^r (see Fig. 2 for such a case). Therefore, we need to take into account also the spanning trees of G_w^r. The crucial observation is, however, that we do not have to handle the potentially exponentially many spanning trees of G_w^r one by one. As we state below, we can recover all information needed to calculate the set of

v-characteristics by combining τ with the set $\mathcal{S}_w^r$ of spanning trees of $\mathcal{B}_w^r$ only. Since G is a d-tenuous graph of bounded cycle degree, the cardinalities of $\mathcal{S}_v^r$ and $\mathcal{S}_w^r$ are polynomial in the size of G. Thus, we need to process only polynomially many combinations of spanning trees from $\mathcal{S}_v^r$ and $\mathcal{S}_w^r$, allowing for a polynomial time algorithm deciding subgraph isomorphism.

We start the formal description of the above considerations by introducing the following notation for the recovery of the subtrees we need from the spanning trees containing w. Let $v \in V(T_G^r)$. Then for any $\tau \in \mathcal{S}_v^r$ and $w \in V(\tau)$ we define $\Theta_{vw}(\tau)$ by

$$\Theta_{vw}(\tau) := \begin{cases} \{\tau \cup \tau' : \tau' \in \mathcal{S}_w^r\} & \text{if } w \in V(T_G^r) \setminus \{v\} \\ \{\tau\} & \text{o/w (i.e., if } w \notin V(T_G^r) \text{ or } v = w), \end{cases}$$

where $\tau \cup \tau'$ is the graph with vertex set $V(\tau) \cup V(\tau')$ and edge set $E(\tau) \cup E(\tau')$. Thus, when $w \in V(T_G^r) \setminus \{v\}$, $\Theta_{vw}(\tau)$ is the set of all trees obtained by "gluing" τ and τ' at vertex w, for all spanning trees τ' of $\mathcal{B}_w^r$, as $V(\tau) \cap V(\tau') = \{w\}$. As an example, the combination of the blue and the red tree in Fig. 2 denotes an element of $\Theta_{vw}(\tau)$. Note that if w is a root vertex different from v then it always has at least one child in $\mathcal{B}_w^r$, that is, τ' is always a tree with at least one edge. In this case, our algorithm needs to combine all pieces of information computed for the children of w in τ, as well as for the children of w in all spanning trees τ' of $\mathcal{B}_w^r$. Otherwise, there are two cases: w is either not a root vertex or it is equal to v. For the first case we have that w has no neighbor outside of $\mathcal{B}_v^r$; for the second case it holds that $\mathcal{B}_w^r = \mathcal{B}_v^r$.

Lemma 2 first considers the case of v-characteristics for subtrees H_u^y for some $y \in \mathcal{N}(u)$, i.e., which can be obtained from H^u by removing the subtree rooted at y (and thus the edge uy as well).

Lemma 2. *An iso-triple (H_u^y, τ, w) of H is a v-characteristic for some $v \in V(T_G^r)$ and $y \in \mathcal{N}(u)$ if and only if there exists a $\theta \in \Theta_{vw}(\tau)$ and an injective function ψ from $\mathcal{N}(u) \setminus \{y\}$ to the children of w in θ such that for all $u' \in \mathcal{N}(u) \setminus \{y\}$ there is a subgraph isomorphism from $H_{u'}^u$ to $(G_v^r|_\theta)_{\psi(u')}^v$ mapping u' to $\psi(u')$.*

A function ψ corresponding to the v-characteristic $(H_u^{u_1}, \tau, w)$ in Lemma 2 is depicted by the dashed lines in Fig. 2. Lemma 3 formulates an analogous characterization for the entire pattern $H = H_u^u$.

Lemma 3. *An iso-triple (H_u^u, τ, w) of H is a v-characteristic for some $v \in V(T_G^r)$ and $y \in \mathcal{N}(u)$ if and only if there exists a $\theta \in \Theta_{vw}(\tau)$ and an injective function ψ from $\mathcal{N}(u)$ to the children of w in θ such that for all $u' \in \mathcal{N}(u)$ there is a subgraph isomorphism from $H_{u'}^u$ to $(G_v^r|_\theta)_{\psi(u')}^v$ mapping u' to $\psi(u')$.*

Lemmas 2 and 3 above are utilized by Subroutine CHARACTERISTICS (see Algorithm 2) in the computation of the set of v-characteristics for a vertex $v \in T_G^r$ with respect to a spanning tree τ of $\mathcal{B}_v^r$ and $w \in V(\tau)$. In particular, for all $\theta \in \Theta_{vw}(\tau)$, we determine the tree τ' satisfying $\theta = \tau \cup \tau'$ (Line 15) and then compute the sets $C_\tau, C_{\tau'}$ of children of w in τ and τ', respectively (Line 16). To decide whether the conditions formulated in Lemmas 2 and 3 for the characterization of v-characteristics hold, we construct a bipartite graph with vertex set $C_\tau \cup C_{\tau'} \dot\cup \mathcal{N}(u)$ and connect a vertex $u' \in \mathcal{N}(u)$ with a vertex $c \in C_\tau$ (resp. $c \in C_{\tau'}$) if and only if there is a subgraph isomorphism from the subtree of H^u rooted at u' to $(G^r_v|_\theta)^v_c$. Note that at this stage, all such partial subgraph isomorphisms have already been computed by the algorithm during the postorder traversal of τ (resp. τ') and stored for all children of w in θ in C_τ (resp. in $C_{\tau'}$). The algorithm then decides whether there exists a matching in the bipartite graph constructed that covers all elements of $\mathcal{N}(u)$ (cf. Lemma 3) or all elements of $\mathcal{N}(u) \setminus \{u'\}$, for all $u' \in \mathcal{N}(u)$ (cf. Lemma 2). The set of v-characteristics with respect to w will then be determined by the outcome of these tests.

3.3 Correctness and Runtime

The correctness of Algorithm 2 follows from Lemmas 2 and 3 by noting that for a leaf $u \in V(H)$ the algorithm finds a v-characteristic (H^y_u, τ, w) for the unique neighbor y of u and for all $\tau \in \mathcal{S}^r_v$, $w \in V(\mathcal{B}^r_v) \setminus \{v\}$, and $v \in V(T^r_G)$. Note that we do not require G to be a d-tenuous outerplanar graph of bounded cycle degree for the algorithm to work correctly. These properties are only used to obtain a polynomial bound on the runtime.

We now claim that Algorithm 2 runs in polynomial time if $G \in \mathcal{O}_{d,c}$. Suppose we call MAIN(H, G) in Algorithm 2 for a tree H with $|V(H)| = k$ and $G \in \mathcal{O}_{d,c}$ with $|V(G)| = n$ for some $k \le n$. Let s be the maximum cardinality of $\mathcal{S}^r_v$ for any $v \in V(T^r_G)$. Then $s \in O\left(n^{c(d+1)}\right)$. Indeed, as G is a d-tenuous outerplanar graph, there are at most $O\left(n^{d+1}\right)$ spanning trees in any block of G. Furthermore, there are at most c v-rooted blocks and an arbitrary number of v-rooted bridges. But bridges have exactly one spanning tree, the bridge itself. The claim then follows by noting that all spanning trees contain v. Since the spanning trees of a graph can be generated with polynomial delay [17], the sets $\mathcal{S}^r_v$ can be computed in polynomial time. Thus, MAIN calls subroutine CHARACTERISTICS a polynomial number of times.

Regarding the complexity of CHARACTERISTICS, we have already seen that there are at most $s \in O\left(n^{c(d+1)}\right)$ spanning trees τ' in $\mathcal{S}^r_w$. The bipartite graph B constructed in Line 17 has at most $|\mathcal{N}(w)| + |\mathcal{N}(u)| \in O(n)$ vertices for every such τ'. Its edges can be constructed by $O\left(n^2\right)$ membership queries to $\mathcal{C}$. We can implement the set $\mathcal{C}$ of characteristics found by the algorithm as a multidimensional array of polynomial size such that each lookup and storage operation can be performed in constant time. A maximum matching on B can be found in $O\left(n^{5/2}\right)$ time [11]. Overall, we compute $O(n)$ matchings for each spanning tree $\tau' \in \mathcal{S}^r_w$ in Lines 18 and 21. (Note, that we can check for the existence of a matching covering all but u' by deleting u' from B.) Thus, CHARACTERISTICS can be implemented to run in polynomial time and the overall runtime of Algorithm 2 is bounded by a polynomial of the size of G.

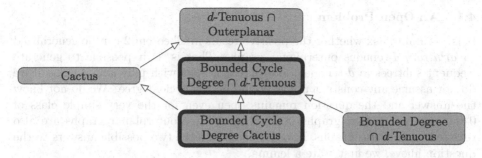

Fig. 3. Results on frequent subtree mining in outerplanar graphs. Green denotes polynomial delay and orange incremental polynomial time pattern mining. Thick borders indicate that the result is new up to our knowledge. Arrows depict subset relations between graph classes (Color figure online).

4   Summary and Open Questions

The main contribution of this work is a polynomial delay algorithm generating frequent subtrees of d-tenuous outerplanar graphs of *bounded* cycle degree. It is based on a polynomial time subgraph isomorphism algorithm that may be of some independent interest. An overview of the results on mining frequent subtrees in different subclasses of outerplanar graphs is given in Fig. 3. Our results (marked with thick border) extend previous results on frequent subtree enumeration (as well as on subtree isomorphism [16]) in outerplanar graphs of bounded vertex degree to a more general class of outerplanar graphs, as bounded cycle degree is less restrictive than bounded vertex degree. Extending the picture with further non-trivial classes is an important challenge for graph mining, not only from theoretical, but also from application aspects.

The algorithm presented in this work is of theoretical interest. As future work, we are going to turn it into a practical algorithm. Our empirical studies on pharmacological molecules show that d and c are both very small numbers for the graphs in this application field.

Another question for future work is whether the positive result of this paper can be generalized to the case that the patterns are d-tenuous outerplanar graphs of bounded cycle degree. In order to calculate the v-characteristics for a root vertex v with respect to a vertex u in the pattern, our algorithm combines at most two sets of spanning trees at any time and assumes that neither u nor the vertices in its local environment are contained in a cycle. Therefore, in order to apply the algorithm to the more general pattern language above, we need the spanning trees of certain local environments of u. However, in contrast to the transaction graphs, it may happen that such spanning trees are composed of the combination of the spanning trees of the blocks for a *non-constant* number of root vertices of the pattern graph. In such a case, an exponential number of spanning trees must be processed. Therefore, if it is possible at all, such a generalization would require a more sophisticated approach.

4.1   An Open Problem

It is natural to ask whether the positive result in Theorem 2 can be generalized to *arbitrary* d-tenuous outerplanar graphs. That is, is it possible to generate frequent subtrees in d-tenuous outerplanar graphs with polynomial delay if we do not assume any constant upper bound on the cycle degree? We do not know the answer and the question remains open even for the very simple class of 0-tenuous outerplanar graphs. Such diagonal-free outerplanar graphs are also referred to as *cactus* graphs. In order to discuss the two possible answers to the question above, we first state a lemma.

Lemma 4. *Let* $\mathcal{P}, \mathcal{G}$ *be graph classes satisfying Conditions 1–3 of Theorem 1, but not Condition 4 , i.e., subgraph isomorphism from* $\mathcal{P}$ *to* $\mathcal{G}$ *is NP-complete. If the FCSM problem for* $\mathcal{P}$ *and* $\mathcal{G}$ *cannot be solved with polynomial delay then* $P \neq NP$.

Proof. Assume that frequent patterns from $\mathcal{P}$ can *not* be generated with polynomial delay in transaction databases from $\mathcal{G}$. If the subgraph isomorphism problem for $\mathcal{P}$ and $\mathcal{G}$ is not in **P**, we are done, as subgraph isomorphism is in **NP**.

Suppose for contradiction that there exists an algorithm $\mathfrak{A}$ that decides in polynomial time for all $P \in \mathcal{P}$ and $G \in \mathcal{G}$ whether P is subgraph isomorphic to G. But then, by using Algorithm 1 with $\mathfrak{A}$ as embedding operator, we get a polynomial delay mining algorithm, contradicting our assumption that such an algorithm does not exist. Thus, there is no polynomial time algorithm deciding whether P is subgraph isomorphic to G and hence $\mathbf{P} \neq \mathbf{NP}$.   □

We are now ready to discuss the open problem formulated above:

(i) Suppose the problem *can* be solved with polynomial delay. Then this would imply that polynomial delay frequent pattern enumeration is possible for NP-complete pattern matching operators, thus solving an open problem.

(ii) Suppose the problem *can not* be solved with polynomial delay. Then, by Lemma 4 above, a proof of this negative result would immediately imply that $\mathbf{P} \neq \mathbf{NP}$. Somewhat surprisingly, deciding subgraph isomorphism from a tree into a cactus graph is an NP-complete problem [2]. Thus Lemma 4 applies to tree patterns and cactus transaction graphs. Therefore, it is likely very difficult to prove for this case that tree patterns can not be mined with polynomial delay.

Thus, both the positive or the negative answer would be a very interesting result. We conjecture that (ii) holds, that is, polynomial delay pattern generation is impossible for computationally intractable pattern matching operators; this is true for graph classes in which the Hamiltonian path problem is NP-complete. If our conjecture holds, then it implies that

(a) in case of intractable pattern matching operators, the primary question should be whether the pattern mining problem at hand can be solved in incremental polynomial time, and not to show that polynomial delay pattern mining is not possible and

(b) *even for very simple graph classes, the cycle degree of the transaction graphs is a crucial parameter for polynomial delay frequent pattern generation.*

References

1. Agrawal, R., Mannila, H., Srikant, R., Toivonen, H., Verkamo, A.I.: Fast discovery of association rules. In: Advances in Knowledge Discovery and Data Mining, pp. 307–328. AAAI/MIT Press (1996)
2. Akutsu, T.: A polynomial time algorithm for finding a largest common subgraph of almost trees of bounded degree. IEICE Trans. Fundam. Electron. Commun. Comput. Sci. **76**(9), 1488–1493 (1993)
3. Chi, Y., Yang, Y., Muntz, R.R.: Indexing and mining free trees. In: ICDM, pp. 509–512. IEEE Computer Society (2003)
4. De Raedt, L.: Logical and Relational Learning. Cognitive Technologies. Springer, Heidelberg (2008)
5. Diestel, R.: Graph Theory, vol. 173. Springer, Heidelberg (2012)
6. Garriga, G.C., Khardon, R., De Raedt, L.: Mining closed patterns in relational, graph and network data. Ann. Math. Artif. Intell. **69**(4), 315–342 (2013)
7. Gottlob, G.: Subsumption and implication. Inf. Process. Lett. **24**(2), 109–111 (1987)
8. Hajiaghayi, M., Nishimura, N.: Subgraph isomorphism, log-bounded fragmentation, and graphs of (locally) bounded treewidth. J. Comput. Syst. Sci. **73**(5), 755–768 (2007)
9. Han, J., Pei, J., Yin, Y., Mao, R.: Mining frequent patterns without candidate generation: a frequent-pattern tree approach. Data Min. Knowl. Discov. **8**(1), 53–87 (2004)
10. Harary, F.: Graph Theory. Addison-Wesley Series in Mathematics. Perseus Books, Boulder (1994)
11. Hopcroft, J.E., Karp, R.M.: An $n^5/2$ algorithm for maximum matchings in bipartite graphs. SIAM J. Comput. **2**(4), 225–231 (1973)
12. Horváth, T., Ramon, J.: Efficient frequent connected subgraph mining in graphs of bounded tree-width. Theor. Comput. Sci. **411**(31–33), 2784–2797 (2010)
13. Horváth, T., Ramon, J., Wrobel, S.: Frequent subgraph mining in outerplanar graphs. Data Min. Knowl. Discov. **21**(3), 472–508 (2010)
14. Horváth, T., Turán, G.: Learning logic programs with structured background knowledge. Artif. Intell. **128**(1–2), 31–97 (2001)
15. Johnson, D.S., Yannakakis, M., Papadimitriou, C.H.: On generating all maximal independent sets. Inf. Process. Lett. **27**(3), 119–123 (1988)
16. Matoušek, J., Thomas, R.: On the complexity of finding iso-and other morphisms for partial k-trees. Discrete Math. **108**(1), 343–364 (1992)
17. Read, R.C., Tarjan, R.: Bound on backtrack algorithms for listing cycles, paths, and spanning trees. Networks **5**, 237–252 (1975)
18. Shamir, R., Tsur, D.: Faster subtree isomorphism. In: Proceedings of the Fifth Israeli Symposium on the Theory of Computing and Systems, pp. 126–131. IEEE (1997)

Author Index

Printed in the United States
By Bookmasters